Advanced Nanoarchitectures with Photocatalytic

Advanced Nanoarchitectures with Photocatalytic Functionality

Zur Erlangung des akademischen Grades eines

DOKTORS DER NATURWISSENSCHAFTEN

(Dr. rer. nat.)

Fakultät für Chemie und Biowissenschaften

Karlsruher Institut für Technologie (KIT) – Universitätsbereich

vorgelegte

DISSERTATION

Von

M.Sc. Ying-Chu Chen

aus

Taipei, Taiwan

KIT-Dekan: Prof. Dr. Reinhard Fischer

Referent: Prof. Dr. Claus Feldmann

Korreferent: Prof. Dr. Dagmar Gerthsen

Tag der mündlichen Prüfung: 06.02.2018

Bibliografische Information der Deutschen Nationalbibliothek

Die Deutsche Nationalbibliothek verzeichnet diese Publikation in der
Deutschen Nationalbibliografie; detaillierte bibliographische Daten sind im Internet
über http://dnb.d-nb.de abrufbar.

1. Aufl. - Göttingen: Cuvillier, 2018

   Zugl.: Karlsruhe (KIT), Univ., Diss., 2018

© CUVILLIER VERLAG, Göttingen 2018

  Nonnenstieg 8, 37075 Göttingen

  Telefon: 0551-54724-0

  Telefax: 0551-54724-21

  www.cuvillier.de

  ISBN 978-3-7369-9780-6

  eISBN 978-3-7369-8780-7

Die vorliegende Arbeit wurde von Oktober 2013 bis Januar 2018 am Institut für Anorganische Chemie des Karlsruher Instituts für Technologie (KIT) unter Anleitung von Prof. Dr. Claus Feldmann angefertigt.

Hiermit versichere ich, die vorliegende Arbeit selbstständig angefertigt und keine anderen als die angegebenen Quellen und Hilfsmittel benutzt zu haben. Wörtlich oder inhaltlich übernommene Stellen sind als solche kenntlich gemacht. Ich habe dabei die Satzung des Karlsruher Instituts für Technologie zur Sicherung der guten wissenschaftlichen Praxis in ihrer gültigen Fassung beachtet.

Abstract

Two novel nanoarchitectures – including the highly branched spikecube exemplified by β-SnWO$_4$ and the biomimetic nanopeapod manifested in Au@Nb@H$_x$K$_{1-x}$NbO$_3$ –were put forward for the first time in this thesis, particularly aiming at enriching the library of pattern designs for sunlight-driven photo(electro)chemical applications.

Specifically, β-SnWO$_4$ spikecubes were entitled on the basis of the peculiar morphology, wherein bundles of nanopillars were self-aligned with quasi-periodicity onto each sharp face of hexahedral cube cores. The reinforced surface roughness stemmes from the nanoarms with a well-defined length(spike)-to-diameter(cube) aspect ratio of 0.3 and gives rise to a nearly 5-fold increase in specific surface area, suggesting more solid surface readily accessible to the reactants to trigger desirable catalytic reactions under illumination of a photoflux. Noteworthily, this geometric engineering was particularly carried out on a Scheelite-type (ABO$_4$) β-SnWO$_4$ crystal in terms of the characteristic twisted coordination structure mediated by the stereoactive electron lone pair, allowing additional textural modifications. Specifically, the interaction between the "non-bonding" electron pair on Sn(II) and the O-2p orbitals tailored the electronic band structure of β-SnWO$_4$ with a visible-light-active band gap of 2.91 eV and a subtile conduction and valence band positions, endowing the photoexcited electron-hole pairs on β-SnWO$_4$ with strong reducing and oxidizing power, respectively. Consequently, an outstanding photocatalytic activity in degrading organic dyes was observed for the β-SnWO$_4$ spikecubes with an enhancement more than 150% in comparison with a benchmark visible-light-active WO$_3$ photocatalyst, which originated mostly from the synergy of the aforementioned effects.

Moreover, the biomimetic peapod-analogous blueprint additionally introduced in this dissertation aims at further subduing photoelectrocatalytic response over a longer wavelength regime far beyond ultraviolet (UV) and blue light. Metal-specific plasmonic nanoantennas were alternatively employed in this design in terms of the straightforward adaptability in light harvesting ability via size, geometry and configuration. Particularly, semi-infinite sub-10 nm Au@Nb core-shell nanoparticle (CS-NP) chain with nanometric breaks residing unidirectionally inside the cavity of tubular H$_x$K$_{1-x}$NbO$_3$ nanoscrolls (NSs). This biomimicry endowes Au@Nb@H$_x$K$_{1-x}$NbO$_3$ nanopeapods (NPPs) with broadband light responses, wherein the niobate NSs and the Au@Nb CS-NP absorb UV and visible light via interband transition and surface plasmon resonance, respectively. More importantly, the strong near-field plasmon coupling between neighboured Au@Nb CS-NPs allows the Au@Nb@H$_x$K$_{1-x}$NbO$_3$ NPPs to absorb near-infrared (NIR) light.

Consequently, the characteristic absorption spectrum matches well with the full solar spectrum. Moreover, the 3D Schottky junction in the NPP structure additionally favors the transfer of plasmon-promoted "hot" electrons from bimetallic Au@Nb to $H_xK_{1-x}NbO_3$ to participate in the subsequent chemical conversion. Overall, a quadruple enhancement relative to an antenna-less photocatalyst and NIR-triggered water splitting as the proof-of-concepts manifest that the Au@Nb@$H_xK_{1-x}NbO_3$ NPPs can readily convert the photon energy into useful chemical fuels.

Zusammenfassung

Zwei neuartige Nanoarchitekturen – darunter β-SnWO$_4$ als hoch verzweigter Spikecube und die biomimetischen „Nanoerbsen", die durch Au@Nb@H$_x$K$_{1-x}$NbO$_3$ repräsentiert werden – wurden in dieser Arbeit zum ersten Mal erwähnt und stellen eine Erweiterung der Bibliothek aus Morphologien für die mit Sonnenlicht angetriebenen photo(elektro)chemischen Anwendungen der.

Insbesondere β-SnWO$_4$-Spikecubes sind aufgrund ihrer eigenartigen Morphologie, bei welcher Bündel aus „Nano-säulen" durch Selbstanordnung regelmäßig auf den Flächen der Würfel gewachsen sind. Die hoch Oberflächenrauheit rührt vom wohldefinierten Länge(Spitze)-zu-Durchmesser(Würfel)-Seitenverhältnis von 0,3 der Nanoarme her und ruft eine nahezu 5-fache Zunahme der spezifischen Oberfläche hervor, was viel Oberfläche andeutet, die für Reaktionspartner zum Initiieren der erwünschten katalytischen Reaktionen unter einer Photoflux-Beleuchtung leicht zugänglich ist. Darüber hinaus wird diese geometrische Besonderheit insbesondere an einem scheeliteartigen (ABO$_4$) β-SnWO$_4$-Kristall hinsichtlich der charakteristisch verdrillten Koordinationsstruktur, welche durch das einzelne stereoaktive Elektronenpaar induziert wird, erzielt. Dies ist die Voraussetzung für zusätzliche strukturelle Modifikationen ermöglicht. Die Wechselwirkung zwischen dem "nicht-bindenden" Elektronenpaar bei Sn(II) und bei den O-2p-Orbitalen verändert die elektronische Bandstruktur des β-SnWO$_4$ zu einer im sichtbaren Licht aktiven Bandlücke von 2,91 eV, was für die Erzeugung Lichtangeregter Elektron-Loch-Paare bei β-SnWO$_4$, sehr hilfreich ist. Daruas folgend konnte eine Erhöhung von mehr als 150% im Vergleich mit einem im sichtbaren Licht aktiven WO$_3$-Photokatalysator als Bezugspunkt festgestellt werden, die von der Synergie der obengenannten Bandstruktur-Effekte herrührt. Folglich wurde an den β-SnWO$_4$-Spikecubes eine herausragende photokatalytische Aktivität beim Abbau der organischen Farbstoffe festgestellt.

Zudem zielt der biomimetische Entwurf einer Erbsenschote, den diese Doktorarbeit beinhaltet, auf eine weitere Adaption der photoelektrokatalytischen Reaktion an eine größere Wellenlänge weit hinter ultravioletten (UV) und blauen Licht. Metallspezifische plasmonische Nanoantennen wurden in diesem Design eingesetzt. Hinsichtlich der unkomplizierten Anpassungsfähigkeit in der Einstellung des Lichtertrages über Größe, Geometrie und Konfiguration sind diese einfach einsetzbar. Insbesondere ist eine Sub-10 nm Au@Nb Core-Shell-Nanopartikelkette (CS-NP) mit nanometrischen Brüchen in eine Richtung im Innern der Kavität der rohrförmigen H$_x$K$_{1-x}$NbO$_3$-Nanoschnecken (NSs) vorhanden. Diese Biomimikry erlaubt den Au@Nb@H$_x$K$_{1-x}$NbO$_3$-Nanopeapods (NPPs) Breitband-Lichtreationen, wobei das Niobat NSs und das Au@Nb CS-NP das UV- und sichtbare Licht durch einen Interband-Übergang bzw. durch die Oberflächenplasmonenresonanz absorbieren. Darüber hinaus ermöglicht die starke Nahfeld-Plasmonenkupplung zwischen dem benachbarten Au@Nb CS-NPs im Au@Nb@$H_xK_{1-x}NbO_3$ NPPs, das nahe Infrarotlicht (NIR) zu absorbieren. Folglich ist das charakteristische Absorptionsspektrum nach dem vollen Sonnenspektrum ausgerichtet. Zudem bevorzugt die 3D-Schottky-Verbindung in der NPP-Struktur den Übergang der plasmonengeförderten "heißen" Elektronen vom bimetallischen Au@Nb zum $H_xK_{1-x}NbO_3$ um diese mit in die nachfolgende chemische Umwandlung einzubinden. Alles in allem zeigt die vierfache Erweiterung eines antennenlosen Photokatalysators zu einer mit NIR initiierten Wasserspaltunge, dass Au@Nb@$H_xK_{1-x}NbO_3$ NPPs die Photonenenergie leicht in nützliche chemische Brennstoffe umwandeln kann.

TABLE OF CONTENTS

1. Introduction and Brief History

Nature is the starting point for advancing science and technology. Replication and adaption of natural systems including elements, structures and processes, so-called biomimicry, routinely help humans to solve problems throughout their existence.[1] One of the extensively studied themes in this field is the photosynthesis in plants. The featured solar-to-chemical energy conversion involved in this course has triggered worldwide scientists and engineers to emulate using man-made materials. This is generally called artificial photosynthesis. The first demonstration was performed in 1839 by Becquerel.[2] In his study, the charge transfer involved in a sunlight-driven chemical reaction was manifested in an electric current flowing from an illuminated silver chloride electrode immersed in an acidic chemical medium to a metallic counter electrode *via* an external circuit (Fig. 1.1). Extensive studies followed up his unprecedented work with a systematic investigation on such photoelectrochemical phenomena in other semiconducting materials, such as Si, Ge, GaAs, ZnO, CdSe, KTaO$_3$, SrTiO$_3$ and TiO$_2$.[3-28] These pioneering works prior to 1970 substantially established the fundamentals of photoelectrochemistry. Afterwards, the work by Fujishima and Honda in 1972 was of particular importance.[29] They used TiO$_2$ as the electrode material, which was irradiated under a photoflux to split water (H$_2$O) and which is similar in many respects to natural photosynthesis. The significance of their results lies in reinforcing the availability of oxygen (O$_2$) and especially of hydrogen (H$_2$), which is the next-generation clean fuel, upon water cleavage under renewable sunlight casting. More importantly, the sustainability and environmental benignity in the context of this photoelectrosynthesis effectively alleviate contemporary public concerns with the oncoming exhaustion of fossil fuel and global temperature increases.

Although the solar fuel generation is of primary importance, other nowadays critical issues in addition to the global warming including food crisis and water stress can also be mitigated *via* artificial photosynthesis. In particular, the reduction of primary greenhouse gas carbon dioxide (CO$_2$) to carbohydrates,[30] the production of ammonium (NH$_4^+$) and nitrate (NO$_3^-$) fertilizers from atmospheric nitrogen (N$_2$)[31] and the remediation of raw and waste water[32] are all possible *via* photoelectrocatalysis, thus stimulating a tremendous research activity in this realm. Further important early contributions emerged between 1978 and 1979 by Bard,[33] who established the principles of photoelectrocatalysis (PEC) not only applicable to a cell configuration but also appropriate for a particulate system (miniaturized metallic counter electrode granulates the semiconductor colloids), and by Nozik,[34] who formulated such concept of a short-circuited photoelectrocatalytic cell as a "photochemical diode".

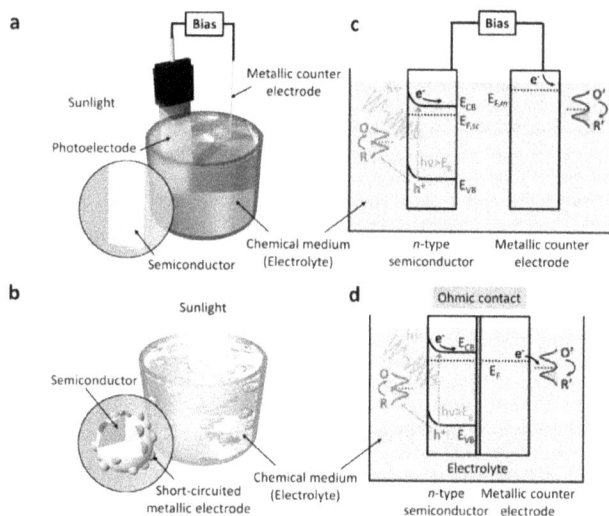

Figure 1. 1. (a) Electrochemical cell and (b) suspended metallized powder configurations for carrying out photoelectrocatalysis using *n-type* semiconductor. Energetics of the (c) photoelectrochemical cell and (d) photochemical diode (framed region in Fig. 1.1b). Abbreviations used: hv, photon energy; E_g, bandgap of the semiconductor; E_{CB}, conduction band edge of the semiconductor; E_{VB}, valence band edge of the semiconductor; $E_{F,sc}$, Fermi level of the semiconductor; $E_{F,m}$, Fermi level of the metallic counter electrode; E_F, Fermi level of the metallized semiconductor.

Their formulations advanced the development in many aspects, including i) the material scope branching out into semiconductors having high electrical resistivity that cannot work as an electrode, ii) the synthetic field opened up to additional more facile but less expensive protocols, and iii) the efficiency record of solar-to-chemical conversion going up to a new plateau owing to strong light scattering within the suspended particles (Fig. 1.2). Later in the early eighties, Grätzel's group excellently exemplified their argumentation with a series of experimental evidences using a variety of semiconducting colloids.[35-37] More importantly, most colloids were characterized by a particle size of few tens of nanometers, rendering these reports acknowledged as another significant milestone highlighting the fusion of modern nanoscience with photoelectrochemistry. The use of nanomaterials in the *bias-free* photochemical diode brings numerous advantages, including shortening the migration pathway of photogenerated carriers (electrons (e⁻) and holes (h⁺)) within semiconducting materials and increasing the surface area for carrier transfer across the solid/liquid interface.[33,34] Most significantly, the specific size quantization effect renders the optoelectronic properties, e.g. the bandgap, the carriers' lifetime and the catalytic activity of these nanoscale granules, size-adaptable.[38-40]

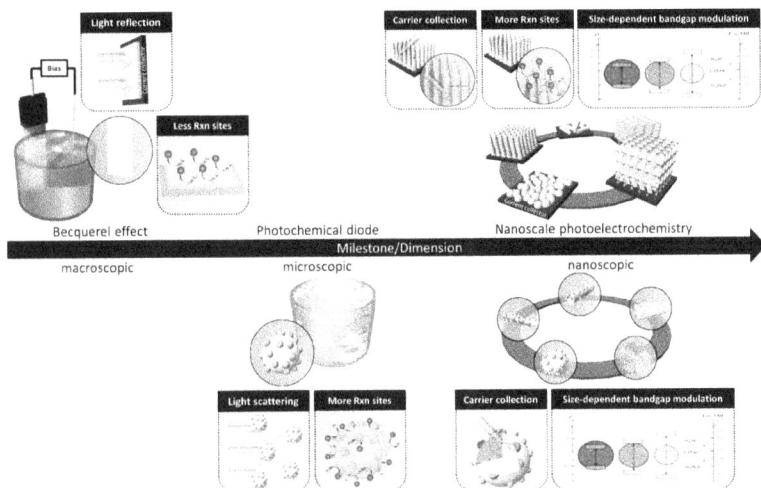

Figure 1. 2. Chronological summary of significant milestones, including dimensional migration and nanostructuring strategies, in photoelectrochemistry. Abbreviations used: L_h, mean free diffusion length of the hole (h$^+$); L_e, mean free diffusion length of the electron (e$^-$). Geometries of the exemplary nanoparticle and nanotip are characterized by the radius (r), the diameter (d) and the height (H).

The intriguing consequences of nanoengineering likewise work well in photoelectrochemical cell and lead to the exploitations in advanced nanoarchitecture that are extensively studied in bilateral classes.[41-44] This study deals with such burgeoning interests via putting forward two modern nanoarchitectures including the highly branched spikecubes exemplified by β-SnWO$_4$ and the biomimetic nanopeapods manifested in Au@Nb@H$_x$K$_{1-x}$NbO$_3$ for the first time to enrich the library of pattern designs for photoelectrocatalysis (Fig. 1.3).[45,46] In-depth discussions on these hierarchical structures begin with an argumentation on the interplay between the textural properties and the local chemical bonding between the constituent elements. The elaborations of the dependency of the architectural geometry on the coordination environment of these chemical systems and the synthetic methodologies follow up to gain a substantial insight into making and especially integrating the nanomaterials at a satisfactory precision level. All in all, this thesis contributes to the synergism of overall atomic and nano-/micro-scopic treatments on the macroscopically photoelectrocatalytic activity.

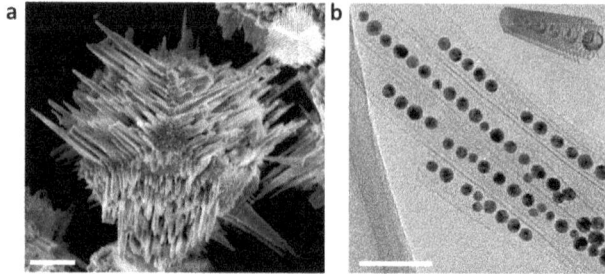

Figure 1. 3. (a) SEM image of multiarmed β-SnWO$_4$ spikecube (scale bar: 2 μm) and (b) TEM image of bioinspired Au@Nb@H$_x$K$_{1-x}$NbO$_3$ nanopeapods (scale bar: 50 nm) (*adapted from reference[45,46]*).

2. Principles of Heterogeneous Photoelectrocatalysis

In order to favor straightforward understanding of most discussions on the photoelectrocatalysis carried out by the topical β-SnWO$_4$ spikecubes and Au@Nb@H$_x$K$_{1-x}$NbO$_3$ nanopeapods artifacts throughout this dissertation, some fundamental nature and basic categories of photoelectrochemistry are first reviewed in this chapter.[33,34] The literal and schematic interpretations mostly deal with *n-type* semiconductors in terms of i) their popularity in the reported literature due to the better stability than that of *p-type* semiconductors and ii) the coincidence that both β-SnWO$_4$ and H$_x$K$_{1-x}$NbO$_3$ are *n-type* semiconductors.

2.1 Review of Energetics

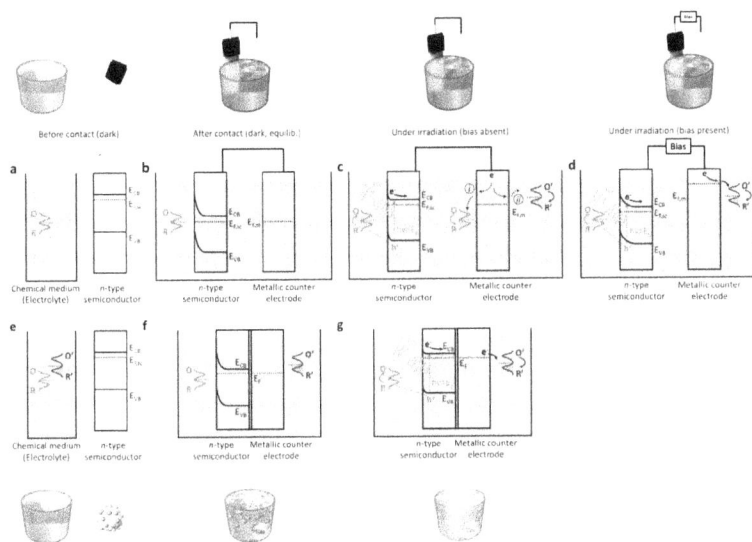

Figure 2. 1. Energetics evolutions in a (a-d) photoelectrochemical cell and (e-g) photochemical diode. (a,d) Energy diagrams of isolated solid and liquid constituents. (b,f) Formation of the semiconductor-electrolyte junction upon mutual contact. Effects of (c,g) suprabandgap light ($hv >$ E_g) illumination and (d) electrical bias on the electronic structure of overall systems. Terms are defined in the text.

All phenomena associated with the "Becquerel effect" started exclusively with the formation of a semiconductor-electrolyte junction at a solid-liquid interface (Fig. 2.1). The semiconductor-electrolyte junction established in the presence of an initial difference in the Fermi level (chemical potential of electrons) between these two phases. The Fermi level in the n-type semiconductor ($E_{F,sc}$) before contact with the electrolyte is located closely below the energy of the conduction band edge (E_{CB}). The Fermi level in the liquid phase is dictated by the presence of redox couples, and the energy is derived from the concentrations and standard potentials. Fig. 2.1a illustrates the conditions of a concurrent presence of the redox couple O and R

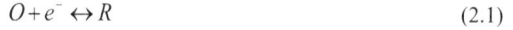

$$O + e^- \leftrightarrow R \tag{2.1}$$

in which the Fermi level is given by

$$E_{F,redox} = V_{redox}^0 + k_B T \ln(\frac{[R]}{[O]}) \tag{2.2}$$

where $V°_{redox}$ (V vs. Normal Hydrogen Electrode (NHE)) is the standard redox potential of the redox couple O/R, k_B the Boltzmann constant, T (K) the absolute temperature and $[R]$ and $[O]$ are their concentrations (mol cm^{-3}), respectively. When the semiconductor is brought into contact with the electrolyte, charge transfer occurs at the boundary until the chemical potential of the electrons ($\overline{\mu_e}$) in the two phases are equivalent (Fermi level equalize). Fig. 2.1b depicts the conditions that the initial Fermi level in a n-type semiconductor is above that of the redox couple O/R in the solution, leading to an electron transfer from the semiconductor to the electrolyte. Upon the acceptance of the electron ($O + e^- \rightarrow R$), the electronic structure of the acceptor (O) changes, particularly, the unoccupied electronic state becomes occupied.

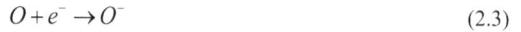

$$O + e^- \rightarrow O^- \tag{2.3}$$

The free energy change associated with this charge uptake step is a measure of the electron affinity (A) of O.

Given the polar nature of the liquid electrolyte, the dipoles in the solvent molecules coordinated to the redox species as a solvation sheath re-orientate in response to this charge variation.

$$O^- \rightarrow R \tag{2.4}$$

Such spontaneous relaxation of the solvation shell results in an additional energy release, the well-known solvent reorganization energy (λ_s). Values for λ_s can be few tenths of an electron volt (eV) up to 2 eV.[47] Moreover, thermal agitation of the solvation structure leads to a Gaussian distribution of the energy states ($D_o(E)$) for the redox species O,

$$D_O(E) = \exp[\frac{-(E - E_{F,redox} - \lambda_s)^2}{4\lambda_s kT}] \qquad (2.5)$$

as depicted in Fig. 2.2. Herein, $E_{F,redox}$ (eV) described in terms of a Fermi distribution function corresponding to the energy state with a 50% probability of electron occupancy[47,48] for the redox couple O/R is

$$E_{F,redox} = -I + \lambda = -A - \lambda \qquad (2.6)$$

I is the ionization potential of the coupled redox species R. Noteworthily, charge transfer occurs only at an overlap of the unoccupied redox states ($D_O(E)$) with occupied semiconductor states. In particular, $D_O(E)$ only becomes zero at energies of $E = \pm\infty$ and is at the maximum when $E = E_{F,redox} + \lambda_s$.

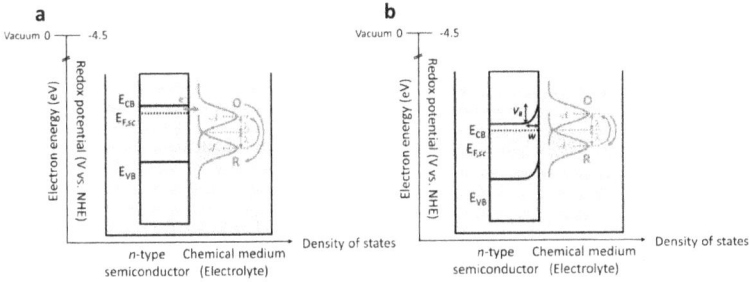

Figure 2. 2. (a) Energy diagram of an *n-type* semiconductor at the initial instant upon contact with the electrolyte with the redox couple O/R. The valence band, conduction band, and the Fermi level ($E_{F,SC}$) of the semiconductor and the distribution of the energy states for the redox species, the Fermi level ($E_{F,redox}$), λ_s, A of the electron acceptor O and the ionization potential (I) of the electron donor R are labeled alongside. Charge transfer carried out at the solid-liquid boundary and reaching (b) the electronic equilibrium at the established semiconductor-electrolyte junction. The developed depletion layer is characterized by the width (w) and the extent of the band bending (V_B).

This interfacial charge transfer results in a semiconductor that is positively charged in regard of the electrolyte. In particular, these charges exclusively distribute within a region adjacent to the interface with the electrolyte, which is designated as "space charge layer" (or "depletion layer" in terms of this space exhausted of majority charge carriers). As a consequence, an electric field is generated and expressed by the bending of the conduction and valence band edges in an upward direction. The extent of band bending (V_B) in the space charge layer is characterized by the difference between $E_{F,sc}$ and V°_{redox}. Specifically, the energy position of the Fermi level is customarily expressed in units of eV with respect to vacuum (zero reference point) in solid state physics. In electrochemistry, the standard redox potential for redox couples is otherwise given in units of volt (V) with respect to NHE as the zero reference point corresponding to the standard redox potential of the proton-hydrogen (H^+/H_2) redox couple. The association between these scales has been corroborated by the effective work function (or Fermi level) of -4.5 eV with respect to vacuum for the standard H^+/H_2 redox couple at equilibrium[49] and is expressed as,

$$E_i \, (eV) = -4.5 - qV_i \, (V \, vs. \, NHE) \tag{2.7}$$

wherein q is the electronic charge. This upward band bending sets up a potential barrier particularly against excess electron transfer into the liquid phase, which is analogous to the rectifying function in the Schottky junction. The width of the depletion layer (w) is determined by,

$$w = \sqrt{2\varepsilon\varepsilon_0 V_B / qN_D} \tag{2.8}$$

wherein ε is the dielectric constant of the semiconductor, ε_0 the permittivity of free space, V_B (V) the degree of band bending, q the electronic charge and N_D (cm^{-3}) is the charge carrier density in the *n-type* semiconductor (equivalent to the density of donors). In typical cases w ranges from 10 nm to several micrometres.[34,47]

Concurrently, on the liquid side charged ions of opposite sign (negative in this case) adsorb onto the surface of the semiconductor for charge offset. In contrast to the semiconductor, the sorption of counter-ions develops only within the well-known Helmholtz double layer bearing a characteristic thickness of a few angstroms (Å) in terms of the carrier density in solution much higher than that in the semiconductor.[33,34,47,48,53] The presence of a Helmholtz layer is of special importance in terms of the electrolyte composition and responsible for the conduction and valence band edges of the semiconductor at the solid-liquid boundary (Fig. 2.3).[53] In other words, polarizing the semiconductor through an artificially applied bias (Fig. 2.1d) doesn't change the relative energy positions of band edges to that of the redox couple in the electrolyte. Such energetic dependency of the band edges exclusively on the chemistry at the phase boundary fairly mirrors another characteristic, namely the Fermi level pinning effect, in a Schottky junction.[34] Altogether, the junctions between semiconductor and liquid electrolytes are in many respects similar to a Schottky junction.

When suprabandgap light is casting at the semiconductor-electrolyte contact (Fig. 2.4), photons having energy hv, where h is the Planck constant and v is the frequency (Hz) of the incident photon, are absorbed by the semiconductor to create electron-hole pairs.

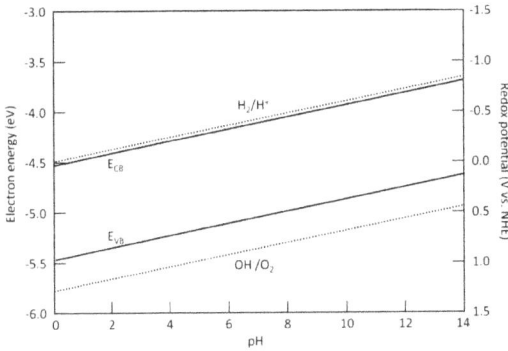

Figure 2. 3. Conduction and valence band edges of a semiconductor oxide plotted as a function of the pH value of coupled aqueous electrolyte. On the pH-E diagram, the E_{CB} and E_{VB} of a semiconductor linearly correlate to the pH value with a slope of 0.059 V/pH at 25 $^{\circ}$C and 1 atm, lying alongside the thermodynamic limits of water electrolysis. This linear dependency is known as the Nernstian relation,[50-53] suggesting that in aqueous solutions H^+ and OH^- are the primary adsorbed species on the surface of a semiconductor within the Helmholtz layer.[50] Two energy scales (eV and V vs. NHE) are shown in parallel for comparison (*adapted from reference [47]*).

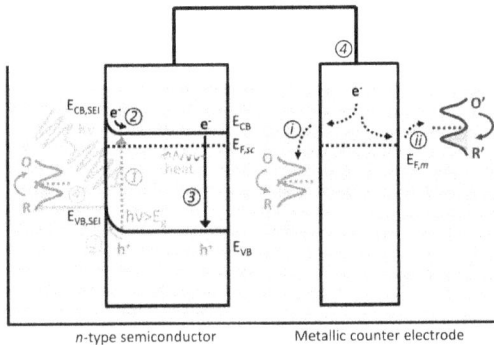

Figure 2. 4. Sequence of main processes in a photoelectrochemical cell, including charge generation upon suprabandgap light irradiation (1), subsequent individual charge transfer (2) or opposite-signed charges recombination (3) and eventual electrode/electrolyte interfacial redox reactions (4). The number of redox couples present in the electrolyte classifies the photoelectrochemical cell into (i) electrochemical photovoltaic cell and (ii) photoelectrocatalytic cell, respectively.

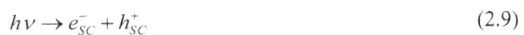

$$hv \rightarrow e_{SC}^- + h_{SC}^+ \qquad (2.9)$$

The electron-hole pairs (excitons) of particular interest are those created in the depletion layer in terms of the built-in electric field resulting in their efficient separation, in which the electrons in the conduction band (CB) move towards the bulk of the *n-type* semiconductor while geminate holes in the valence band (VB) migrate to the solid-liquid phase boundary.

$$e^-_{depletionlayer} \rightarrow e^-_{bulk} \tag{2.10}$$

$$h^+_{depletionlayer} \rightarrow h^+_{semiconductor-liquid\,junction} \tag{2.11}$$

Most excitons photogenerated beyond the space charge layer suffer from recombination,

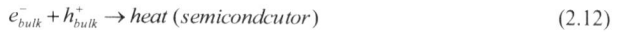

$$e^-_{bulk} + h^+_{bulk} \rightarrow heat\,(semicondcutor) \tag{2.12}$$

except for those that successfully escape from that process by virtue of the minority holes diffusion into the depletion layer. The photogeneration and subsequent separation of electron-hole pairs in the depletion layer brings about $E_{F,SC}$ arising to the pristine energy level (Fig. 2.1c), namely the energy position prior to the initialization of charge transfer at the semiconductor-electrolyte junction (Fig. 2.1a). This original energy status corresponds to a measure of the semiconductor potential with respect to NHE, wherein no excess charge is present in the semiconductor and which is designated as the "flatband" potential (V_{fb}). The holes effectively scavenged to the solid-liquid phase boundary react with redox specie R in the electrolyte at a potential corresponding to the band edge of VB at the semiconductor-electrolyte interface ($E_{V,SEI}$), leading to the oxidation of R to O.

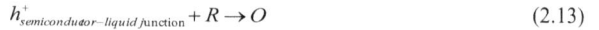

$$h^+_{semiconductor-liquid\,junction} + R \rightarrow O \tag{2.13}$$

More or less, the charge carriers may trigger undesirable self-oxidation of the semiconductor depending on the thermodynamic redox potentials of the electrode decomposition reaction (Fig. 2.5).[33,54]

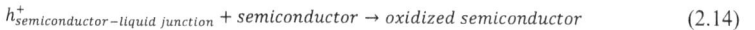

$$h^+_{semiconductor-liquid\,junction} + semiconductor \rightarrow oxidized\,semiconductor \tag{2.14}$$

Concurrently, the majority electrons delivered toward the semiconductor bulk are subsequently shuttled from the semiconductor *via* an external circuit to the metallic counter electrode and thereat injected into the electrolyte to drive a reduction reaction.

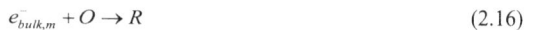

$$e^-_{bulk,SC} \rightarrow e^-_{bulk,m} \tag{2.15}$$

$$e^-_{bulk,m} + O \rightarrow R \tag{2.16}$$

Specifically, the most negative potential available for this photoexicted electron is V_{fb}. Two distinct photoelectrochemical cells, i) electrochemical photovoltaic cell and ii) photoelectrocatalytic cell, are generated based on the nature of electrolyte (Fig. 2.4).[33,34,53]

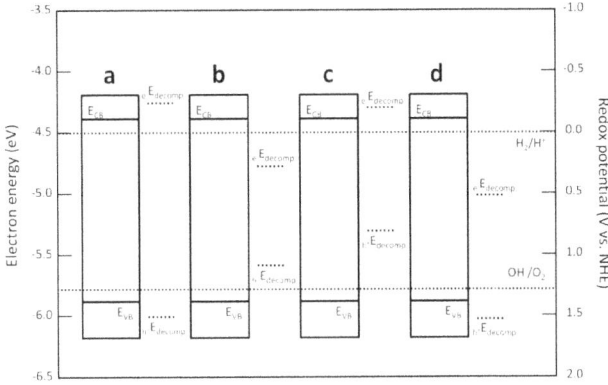

Figure 2. 5. Thermodynamics regarding photodecomposition of semiconductors. The cathodic reduction and anodic oxidation of semiconductors are always associated with the photogenerated electrons in the CB and geminate holes in the VB as the electronic reactants. The Fermi energies related to these redox degradations are characterized as $_{h^+}E_{decomp.}$ and $_{e^-}E_{decomp.}$, respectively. The thermodynamic stability of a semiconductor/electrolyte system depends decisively on the relative position of the decomposition Fermi levels ($_{h^+}E_{decomp.}$ and $_{e^-}E_{decomp.}$) to the conduction and valence band edges (E_{CB} and E_{VB}) of a semiconductor at the solid-liquid boundary. The semiconductor is (a) thermodynamically steady, provided that $_{h^+}E_{decomp.}$ is more positive (on scale of V vs. NHE) than E_{VB} and $_{e^-}E_{decomp.}$ is more negative (on scale of V vs. NHE) than E_{CB}. In contrast, semiconductors become (b) thermodynamically unsteady at inverse conditions. Most *n-type* semiconductors studied to date were either (c) cathodically or (d) anodically stable (*adapted from reference [54]*).

In electrochemical photovoltaic cells, the electrolyte comprises one effective redox couple O/R only and the oxidation reaction (Eqn. 2.13) at the semiconductor side is entirely reversed at the metal electrode side (Eqn. 2.16). In other words, no net variation in the composition of the electrolyte occurs. In this mode, the incident photon energy is thoroughly converted into electrical work in the ideal case, rendering photocurrent and photovoltage for usage. In the photoelectrocatalytic cell, multiple redox couples are concurrently present in the electrolyte, leading to distinct oxidation (Eqn. 2.13) and reduction reactions

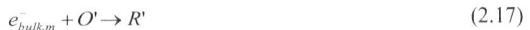

$$e^-_{bulk,m} + O' \rightarrow R' \tag{2.17}$$

11

at the anode and cathode, respectively. Consequently, a chemical change of the electrolyte takes place. Noteworthily, the radiant energy in this system is either employed to overcome the activation barrier of the overall cell reaction only or further converted into the chemical free energy of the substances generated from the electrolyte (Fig. 2.6).

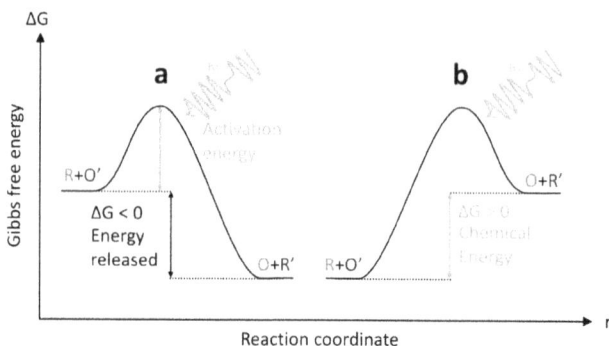

Figure 2. 6. (a) In the photoelectrocatalytic cell, an exergonic ($\Delta G < 0$) reaction (R + O'→ O + R') is accelerated in the presence of the illuminated semiconductor electrode. (b) In the photoelectrosynthetic cell, an endergonic ($\Delta G > 0$) reaction is triggered in the presence of an illuminated semiconductor electrode, leading to the conversion of radiant to chemical energy.

In this regard, the photoelectrocatalytic cell that can store the photon energy in chemical fuels is particularly specialized and designated as "photoelectrosynthetic cell". In particular, the photoelectrosynthetic cell is characterized by a cell reaction with a positive free energy change ($\Delta G > 0$).

The principles of photoelectrosynthetic and photoelectrocatalytic cells can be extended to the design of particulate systems (Fig. 2.1e-g). Such configuration was labeled as "photochemical diode", made of a photoelectrocatalytic or photoelectrosynthetic cell collapsing into monolithic particles without an external circuit. For instance, a particulate metal electrode fused into a powdered *n-type* semiconductor forming ohmic contacts between is the prototype of a Schottky-type photochemical diode (Fig. 2.1e). In analogy to the cell counterpart, the reaction is carried out by simply immersing the photochemical diodes in an electrolyte containing the redox couples (Fig. 2.1f) with the semiconductor faces illuminated (Fig. 2.1g). The photoprocess (Eqn. 2.13), e.g. photooxidation of organic solutes or water solvents, is performed at the irradiated semiconductor side while the concomitant reduction (Eqn. 2.17) of dissolved O_2 to $O_2 \cdot^-$ or water to H_2 proceeds at the dark metal site.

The dimension of a photochemical diode is arbitrary, in which microscale particles ensure adequate light harvesting (Fig. 2.7a), and the depletion layer is fully-developed (Fig. 2.7b), as long as the particle diameter (d) matches the light penetration depth (α^{-1}). It is larger than the width of space charge layer (w). Alternatively, nanoparticles render optimal charge collection (Fig. 2.7c), provided that the particle size is in the same range of the mean free diffusion length of hole/electron (L_h/L_e). Additionally, nanoparticles offer abundant solid/liquid interfaces, allowing numerous charges available to the redox couples to carry out certain reactions, such as sluggish water oxidation with four-electrons/four-protons involved (Fig. 2.7d).

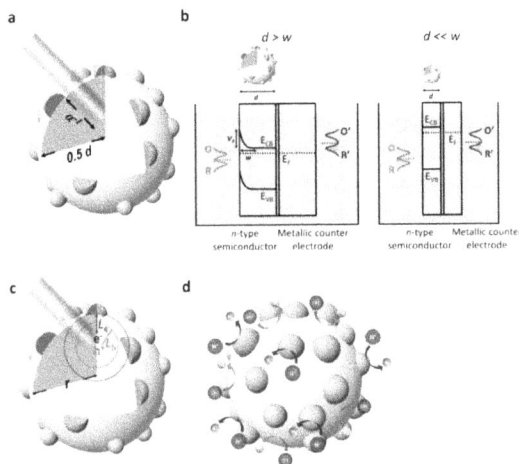

$$2H_2O \rightarrow O_2 + 4H^+ + 4e^- \qquad (2.18)$$

Figure 2. 7. Effects of particle size in (a) light absorption, (b) development of space charge layer, (c) charge collection and (d) interfacial charge transfer rate for photochemical diode. More than 90% of the incident light is absorbed by the photochemical diode when the particle diameter is more than 2.3-fold the value of α^{-1}.[44,55] The space charge layer is instantaneously formed in a microscale photochemical diode but gradually degenerates upon downscaling the dimension to the nanoscale. When the particle size is smaller than the number of w, the conduction and valence bands are essentially flat, which can hardly separate the photogenerated electron/hole pairs. Ideal charge collection is yet feasible when the particle diameter is further reduced to the same scale of L_e/L_h. The characteristic high surface area of nanoscale photochemical diodes promotes a charge transfer rate to perform reactions with slow kinetics. Abbreviations used: α, wavelength-dependent absorption coefficient (cm^{-1}); r, particle radius.

The pros and cons of the particle dimension call for engineering a hierarchical photoelectrocatalyst, integrating ingredients at different length scales into an elaborate architecture for desired functionalities.[41-44] This places exceptional demands on systematics of material property and preparation, characterization and analysis, and particularly the interplay between micro- and nanoscale components, as illustrated in the sections below.

2.2 Systematics of Elements Constructing Semiconducting Photoelectrodes and Particulates as Photoelectrocatalyst

Figure 2. 8. Elements making up photoelectrocatalytic materials (*adapted from reference [41]*).

The electronic structure of a semiconductor is featured by the presence of a bandgap, namely an energy interval with very few electronic states between the conduction band and the valence band having high densities of states.[56] In the context of photoelectrocatalysis, identifying the highest occupied and lowest unoccupied electronic levels in the semiconductor is of great importance since they primarily dictate the charge transfer at the semiconductor-electrolyte junction. Particularly, the highest occupied energy levels (HOMO) corresponding to the top of valence band (E_{VB}) are a measure of the ionization potential (I) of the bulk material. The lowest unoccupied energy levels (LUMO) corresponding to the bottom of conduction band (E_{CB}) are a measure of the electron affinity (A) of the bulk material.

On this basis, Butler and Ginley formulated an empirical evaluation of the band edges.[50] In the argumentations, the band edges can be derived from the bulk electronegativity (χ) of a material, which is the geometric mean of the electronegativities of the constituent elements. Moreover, their study manifested that the calculated band edges were in good agreement with those extracted from experimental measurements, e.g. the most reliable Schottky-Mott method.[34,48]

On such basis, elements are categorized into four groups responsible for forming i) crystal as well as electronic structures, ii) crystal structure alone, iii) impurity levels, and iv) additional co-catalyst, respectively (Fig. 2.8). For non-transition metal oxides, sulfides and nitrides bearing metal ions with d^0 and d^{10} configurations (e.g. Ti^{4+} and Zn^{4+}), the conduction band consists mostly of the metal d and sp orbitals, respectively.[47,57] The valence bands are otherwise composed of O $2p$, S $3p$ and N $2p$ orbitals for these metal oxides, sulfides and nitrides, respectively.[41,44,47,58] The valence band edges in non-transition metal oxide photocatalysts are located at energy levels approximating the absolute electronegativity of oxygen (-7.54 eV), endowing the photogenerated holes in the valence band with strong oxidizing power. However, the conduction band edges of these oxides are otherwise close to the reduction potential of H_2/H^+ redox couple, situating far from the valence band edges. Consequently, most have large bandgaps and respond only to ultraviolet (UV) light.

In contrast, the valence band edges in non-transition metal nitrides and sulfides are in general more negative (on the redox potential scale) than those of oxides in terms of the smaller electronegativities (on the electron energy scale) of nitrogen (-7.30 eV) and sulfur (-6.22 eV). As a result, most nitrides and sulfides have favorable bandgaps, allowing the electrons to be excited by the visible light, whereas the created holes in the valence band have minor oxidizing power. Likewise, the discrepancy in electronegativity between nitrogen, sulfur and oxygen renders these non-oxides prone to anodic photocorrosion in aqueous electrolytes.[41,54]

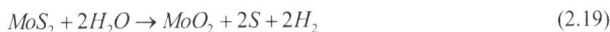

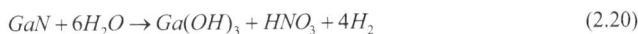

$$MoS_2 + 2H_2O \rightarrow MoO_2 + 2S + 2H_2 \qquad (2.19)$$

$$GaN + 6H_2O \rightarrow Ga(OH)_3 + HNO_3 + 4H_2 \qquad (2.20)$$

The instability limits their employment to photoelectrocatalysis in the presence of sacrificial reagents (e.g. S^{2-} and SO_3^{2-}) and passivation layers.[41,44] In the photoelectrocatalysis community, the basic scientific focus on them is particularly placed on visible light sensitization of metal oxide photoelectrochemical cells or photochemical diodes,[59] interfacial charge transfer,[60] and quantum confinement effects.[61,62] With respect to the visible light sensitization, orbitals of Pb $6s$ in Pb^{2+}, Bi $6s$ in Bi^{3+}, Sn $5s$ in Sn^{2+} and Ag $4d$ in Ag^+ can likewise form valence bands above that of O $2p$ orbitals in metal oxide photocatalysts.[41] The valence band engineering is dictated by the quantity of these metal cations incorporated into the parental metal oxides and the eventually formed crystal structure.

Additionally, doping also plays an important role in tailoring the band structure.[41,63] Unlike the valence band treatment, discrete electronic states are created in the band gap of the metal oxide host. Transition metal cations with partially filled d orbitals (e.g. Cr^{3+}, Ni^{2+} and Rh^{3+}) are effective dopants to endow the metal oxide photocatalysts with visible light response.[41]

In contrast, orbitals of alkali, alkaline earth and lanthanide ions do not participate in the electronic structure adaption and merely contribute to the crystal structure formation. Noble metals (e.g. Au, Pt and Rh) and some transition metal oxides (e.g. NiO, RuO_2, IrO_2) function as co-catalysts to favor the metal oxide photocatalysts to readily tackle challenging photoinduced processes. The presence of co-catalysts at the semiconductor-electrolyte junction not only relaxes the activation barrier of the reaction but also allows spatial separation of photogenerated electron-hole pairs.[41,44]

2.3 Toolbox for Realizing Material Design and Nano-origami of Photoeletrocatalysts

Extensive pioneering works on the photoelectrochemistry of simple unitary and binary semiconductors date back to the early fifties with Ge, TiO_2 and CdS as prototypical examples.[4-11] Virtually, the studies on nonoxidic photoelectrocatalysts are of a rather fundamental nature in terms of the aforementioned photocorrosion issue. Although the TiO_2 photoelectrocatalyst is of longstanding practical interest in terms of its stability, low cost and nontoxicity, superior efficiency is reported particularly for photocatalytic downhill reactions (in terms of Gibbs free energy change) in particulate photochemical diode configurations.[64-66] Contrarily, additional electrical (external voltage) or chemical (pH difference between catholyte and anolyte) biases are indispensable to perform a photoelectrosynthetic uphill reaction owing to the inferior E_{CB} of TiO_2.[65-67] Consequently, photoelectrosynthetic water splitting and carbon dioxide reduction are mostly carried out in cell configuration.[29,30,67] Such obsession can be circumvented by conduction band adjustment *via* crystallographic control.

For instance, TiO_2 occurs naturally in three polymorphs (Fig. 2.9), namely rutile (tetragonal, space group $P4_2/mnm$), anatase (tetragonal, space group $I4_1/amd$) and brookite (orthorhombic, space group *Pbca*). As a bulk material (particle size > 35 nm), rutile is the most stable phase with E_{CB} coinciding with the thermodynamic potential (0V vs. NHE) for proton reduction (H^+/H_2).[65] By contrast, brookite and anatase turn out to be thermodynamically favored in TiO_2 nanocrystals with the dimension ranged from 11 to 35 nm and less than 11 nm, respectively. The potential of the conduction band electrons in the anatase-TiO_2 is more negative (on the redox potential scale) by approximately 200 mV than that of rutile. Among these polymorphs, brookite has the most negative

E_{CB} that allows photoelectrosynthesis most likely carried out in the simplest photochemical diode scheme. Such phase-dependent energetics is likewise adopted by other chemical systems including iron oxides, molybdenum disulfides and so forth.[68,69]

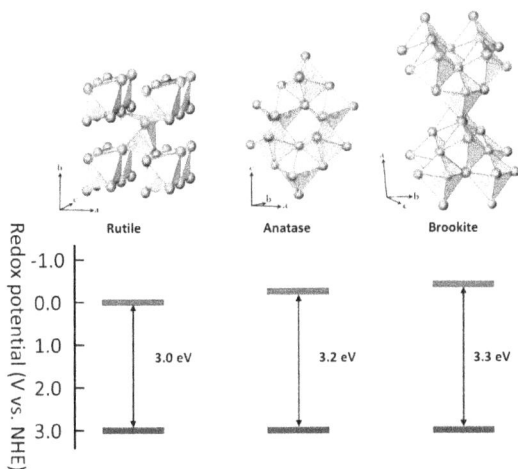

Figure 2. 9. Crystal and band structures of TiO$_2$ polymorphs at pH = 0 (*adapted from reference [65-67]*).

The methodology for nanocrystal preparation herein is grouped based on the phase of the reaction reservoir, wherein the nanoparticles are formed. Particularly, the great versatility in solution- and vapor-phase processes renders these methods (e.g. hydrothermal and polyol-mediated approaches) that are known for the synthesis of pure anatase,[65,71] brookite[65,70] and biphasial anatase/brookite colloids.[65,71] Moreover, excellent manipulation of the geometry, e.g. morphology, crystal facets and size, is manifested in the derived nanocrystals, which further fine tunes the electronic structure to fit specific functional demands.

Alternatively, the band structure of TiO$_2$ can likewise be reframed *via* developing new titanates with multinary composition.[41,63] For example, ternary SrTiO$_3$ and quaternary K$_2$La$_2$Ti$_3$O$_{10}$ demonstrate a more negative E_{CB} position (on the redox potential scale) than that of parental TiO$_2$, allowing these materials to perform photosynthetic uphill reactions in particulate configuration.[41] Efficient photoelectrocatalysts are additionally available in other chemical systems including niobates (e.g. K$_4$Nb$_6$O$_{17}$ and Ca$_2$Nb$_2$O$_7$) and tantalates (e.g. NaTaO$_3$ and Ba$_4$Ta$_4$O$_{15}$). Noteworthily, these multicomponent oxidic photoelectrocatalysts are mostly layered perovskite structures in terms of the versatility to accommodate most of the metallic ions in the periodic table together with a significant number of other anions (Fig. 2.10).[63]

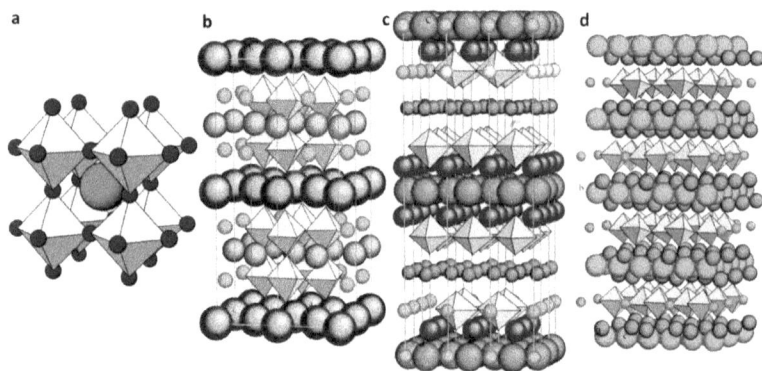

Figure 2. 10. (a) Ideal ABO_3 cubic Perovskite structure (octahedral BO_6 units with A cation located in the middle of the cubic cage) and other perovskite-related layered variants including (b) Dion-Jacobson phases with a general formula $A_{n-1}B_nO_{3n+1}$ typified herein by $RbLaNb_2O_7$ (yellow, NbO_6 units; green, Rb atoms; lilac, La atoms), (c) Ruddlesden-Popper phases with a general formula $A_{n+1}B_nO_{3n+1}$ exemplified herein by $Li_2CaTa_2O_7$ (yellow, TaO_6 units; blue, Ca atoms; red, Li atoms; green, O atoms) and (d) Aurivillius phases with a general formula $(A_{n-1}B_nO_{3n+1})^{2-}$ illustrated herein by Bi_2WO_6 (yellow, WO_6 units; lilac, Bi atoms; green, O atoms). In all cases, n represents the number of the perovskite-like layers (*adapted from reference [63]*).

The most straightforward synthesis employs solid-phase techniques (viz. solid state reaction), wherein parental metal oxides and alkali/alkaline earth carbonates as starting materials are calcined in air at high temperature.[41] Most oxides produced inthis way are well-crystalline, which favors the migration of photogenerated charges. The formation of large particles upon sintering supports the development of space charge layer to boost the separation of charge carriers. These effects are advantageous for the photoelectrocatalytic activity. On the other hand, giant particle size results in long delivery distances, which increase the recombination probability of electrons and holes, and low surface area, which impairs the adsorption of redox couples at the solid-electrolyte interface. These effects are disvantageous and thereof call for nanoengineering, wherein soft chemistry again offers a powerful toolbox.

The simplest scheme employs a soft-chemical transformation of oxides preformed in the solid state reaction to generate a variety of nanoarchitectures that retain many of the textural features of the precursor phase. To preserve most of the bond connectivity in the parental structure, reactions must be carried out at a sufficiently low temperature. Many of these reactions involve ion-exchange, intercalation, and oxidation-reduction chemistry at discrete reaction sites within the solid.[72,73]

In particular, ion exchange is the most common dimensional reduction formalism for layered pervoskites.[72] The interlayer cations or structural units, which serve as a "bridge" to form bonds between basal lattice planes, can readily be interchanged at low temperature. For instance, the mobile interlayer K^+ cations in the layered $K_2La_2Ti_3O_{10}$ can be exchanged by protons from warm strong acid (e.g. H^+ in HCl) to form $K_{2-x}H_xLa_2Ti_3O_{10}$ (Fig. 2.11). This protonated titanate is a solid acid that can intercalate a variety of organic bases, such as the sterically bulky tetra-(n-butyl)ammonium hydroxide.[72] Consequently, the interlayer gallery opens up and swells with solvent until the forces that hold the layers together are overcome. Eventually, the solid delaminates to form a colloidal suspension consisting of molecular sheets and/or tubes.[72,74]

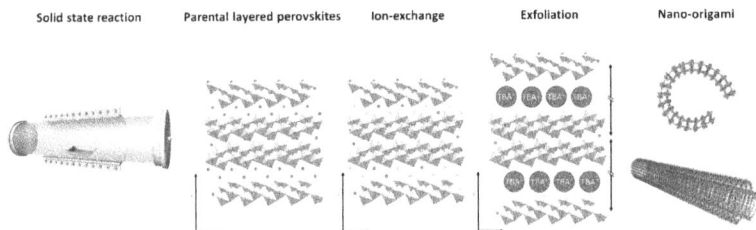

Figure 2. 11. Schematic illustration of stepwise nano-origami of multinary layered perovskite preformed during the initial solid-state reaction, and subsequent ion-exchange and scrolling to generate texture-preserved tubular nanocrystals.

Specifically, these exfoliated perovskite colloids are nanoscale and monocrystalline, leading to shortened and fluent transport of photogenerated charges and rendering them readily available to the redox couples due to enhanced surface areas.[44,72] Similar efficacy is yet observed in the self-aggregate, as evidenced by the agglomerate of $HCa_2Nb_3O_{10}$ nanosheets that clearly outperform the non-exfoliated condensed counterpart in photoelectrocatalytic H_2 generation by up to one order of magnitude.[75,76] Such conclusion stimulates the prevalence of this soft-chemical multistep process (colloidal synthesis and subsequent integration) for the fabrication of photoelectrocatalysts serving as either the photochemical diode or the photoelectrode in the cell system.[42] Most importantly, additional flexibility is propoble in the configuring process, including hierarchical architecture formation, and moreover, the incorporation of additional guest nanocrystallines into the host framework, to further refine the quality of the consequent nano-assemblies.

The feasibility of soft-chemically merging diverse nanomaterials opens up a new processing route complementary to the preceding strategies, aiming mostly at endowing those individual parental constituents with additional functionalities. Particularly, the broadband light harvesting and charge rectification abilities are of primary interest.[42,44] The former originates from the inevitable trade-off in the aforementioned energetics engineering between the redox power and the light absorption capability. For instance, the overall Gibbs free energy change of approximately 2 eV specific to the overall photoelectrosynthetic water splitting establishes the bandgap threshold of the semiconducting photoelectrocatalyst employed solitarily to trigger this reaction.[77] Nevertheless, such limitation breaks down upon decoupling the sub-tasks of light absorption from water electrolysis. To this end, molecular organic dyes, semiconducting quantum dots (QDs) and plasmonic metal nanoantennas are highly appealing photosensitizers.[78-80] Moreover, the heterojunction formed between the host semiconductor and guest sensitizer allows the inter-component transfer of photogenerated charge carriers while the electric field developed at the interface effectively suppresses the backflow and recombination. This concomitant charge rectifying characteristics is of great importance to the bias-free photochemical diode in view of the undesirable charge loss thereof readily quenched.[44,81,82]

In addition to these synergies, exceptional collective properties stemming from the inter-entity coupling can be introduced to the nanocomposites *via* engineering the spatial arrangement with nanoscale resolution over micro- and macroscopic distances.[83-86] For instance, when the surface electrons (excitons) in adjacent nanocrystalline semiconductors are electronically coupled, new bonding and antibonding states are formed at the interface, endowing the nano-assembly with a reframed energy band. Such short-range phenomenon has been reported for clusters of anatase-TiO_2 nanocrystals and paired CdTe nanoparticles separated by less than 2 nm, giving rise to a much narrower bandgap of these nano-assemblages than that of the isolated counterparts.[84,85] Virtually, the electronic coupling can be found in the nano-assemblies consisted of either semiconducting or metallic nanoparticles and in the nanocomposites (Fig. 2.12). Moreover, such interplay becomes profound in large-scale arrays owing to additional aid resulting from long-range dipole-dipole interactions.[43] The bandgap narrowing allows those nano-systems to more efficiently take advantage of longer wavelength light in the solar spectrum for photoelectrocatalytic use, which urges tremendous research interests in hierarchical composite nanostructures with well-organized patterns.

Given the electronic coupling valid only at few nanometres scale, conventional top-down photolithography, however, turns out to be highly challenging for such nano-origami in terms of practical and theoretical limits of the resolution.[87,88] Contrarily, bottom-up soft chemistry offers breakthroughs to these confinements.[89-96] In consequence, a rich library of one-, two- and three-dimensional nanoarchitecutres with precisely controlled inter-particle distances and ordering over multiple length scale was generated by such fabrication platforms including template-assisted, molecular-linker-mediated (e.g. DNA and surfactant molecules) and block-copolymer-based self-assemblies.[97-100]

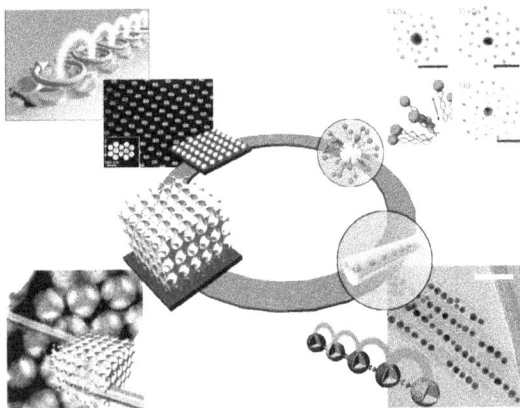

Figure 2. 12. Exemplary nano-assemblies including naphthalene-analogous planar arrays (upper left corner), core-satellite (upper right corner) and biomimetic peapod (lower right corner) superstructures that allow the electronic coupling of surface plasmons on metallic building blocks, and inverse opal architecture (lower left corner) that allows the optical coupling between incident light and the solid photonic crystal, respectively (*adapted from reference [46,97-99]*).

3. Analytical Tools for Photoelectrocatalyst Characterization

Evidently, burgeoning progresses are ongoing in the exploitation of promising photoelectrocatalysts and versatile toolkits for additional optimization *via* nanoengineering and integration. In response to such proliferation, the characterization metrics to give systematic insights into the material, the treatment and the interplay with photoelectrochemical performance are streamlined to boost the advances further.[42] The following section discusses the measurements, which are of central importance to the photoelectrocatalyst, and moreover, particularly performed on the topical β-SnWO$_4$ spikecubes and Au@Nb@H$_x$K$_{1-x}$NbO$_3$ nanopeapods of this monograph, in detail.

3.1 X-ray Powder Diffraction (XRD)

X-rays with energies ranging from ca. 100 eV to 10 MeV are classified as electromagnetic waves with featured wavelengths corresponding to ca. 10 to 10^{-3} nm.[101] Given such wavelengths comparable to the spacings between atoms in a crystal, X-rays are preferentially employed to obtain atomic scale information of materials.[102-106] Such knowledge is highly prized, provided that the atomic structure of chemical systems intimately defines the ultimate macroscopic properties.[107] Particularly, X-ray diffraction crystallography for powder samples is a highly mature discipline that is prevalently deployed to probe the spatial arrangement of atoms in a variety of substances.[102-106] Virtually, only X-rays with wavelengths in the range of a few angstroms (Å) to 10^{-1} Å (viz. hard X-ray with energies of ca 1 to 120 KeV) are used for this application.

X-rays are generated upon rapid deceleration of high-speed charged particles (typically electrons) with high kinetic energy.[102,105,106] Such process is carried out in a X-ray tube, which is the primary X-ray source in laboratory X-ray diffractometers (Fig. 3.1), *via* applying a high voltage up to several tens of kV between two electrodes. Consequently, the electrons are extracted out from the cathode and accelerated across the voltage field toward the anode (metallic target). When the electron beam bombards the target, interacts with the atoms therein (deflection) and slows down, X-rays are radiated from the anode with the energy derived from the kinetic energy loss of the incoming electrons. Given the decelerated patterns varied with electrons, continuous X-rays with various wavelengths, the well-known *Bremsstrahlung*, are generated. Particularly, when the electrons give up all the kinetic energy in a single inelastic deflection, the consequently originating X-rays have the maximum energy (E_0) and shortest wavelength (λ_{SWL}) that can be estimated from the accelerating voltage (V) between two electrodes,

$$E_0 = qV \tag{3.1}$$

$$\lambda_{SWL} = \frac{hc}{qV} \tag{3.2}$$

wherein q the electronic charge, h the Planck constant and c is the speed of light.

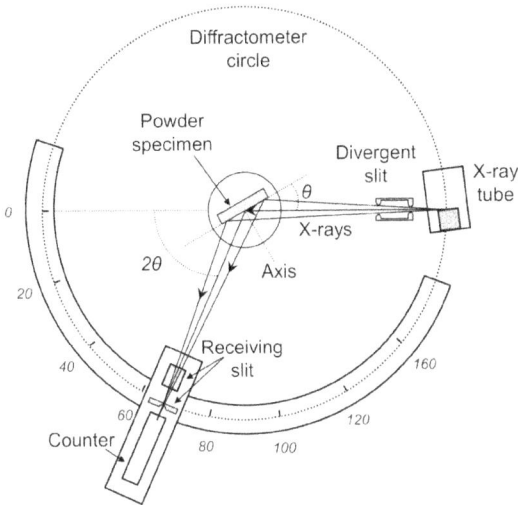

Figure 3. 1. Basic concept of an X-ray diffractometer (*adapted from reference [105]*).

Noteworthily, beyond a certain voltage threshold (viz. the excitation potential), the incident electrons have sufficient kinetic energy that can be transferred to the core electron of a target atom to allow the ejection from the inmost K shell (*1s* orbital). In consequence, the atom becomes excited due to the created core hole, which is then restored to the stable state upon refilling this vacancy by the outer shell electron (e.g. L or M shells) having more positive energy. Excess energy is given out in such process upon emitting discrete *characteristic X-rays* bearing featured wavelengths (energies) specific to the target metal. Standard descriptive nomenclature for the *characteristic X-ray* depends on the electron shells performing such electronic transition. For instance, the transitions between shells $L \rightarrow K$ or $M \rightarrow K$ emanate K_α and K_β X-rays, respectively. Given the XRD measurement required monochromatic X-ray, the presence of these strong and sharp *characteristic X-rays* renders the study of crystal structure feasible. The most popular target materials implanted in the X-ray tubes include copper (Cu) and molybdenum (Mo) yielding nearly monochromatic K_β lines bearing

23

wavelengths of 1.54 and 0.71 Å. To extract these K lines superimposed by the *Bremsstrahlung*, a filter made of materials having an atomic number (Z) one or two less than that of target metal with the X-ray absorption edge somewhat preceding the wavelength of the *characteristic X-rays* is inserted into the incident ray path. Moreover, a crystal monochromator in conjunction with the X-ray diffractometer can further reinforce the monochromatism of the *characteristic X-rays*.

When monochromatic X-rays encounter the powder specimen under investigation, the oscillating electric field of this electromagnetic wave primarily brings the charged electron in preference to the nucleus of the material atom into a sinusoidal vibrating motion. Consequently, the electron is periodically accelerated and decelerated, and hence emits new X-rays having equivalent wavelength and energy to the incident radiation. In this sense, X-rays are elastically scattered by the electron (Thomson scattering) and the scattered beam is *coherent* with the incoming ray. Virtually, diffraction is simply a scattering phenomenon, wherein appreciable coherently scattered X-rays mutually reinforce one another, giving rise to a set of diffracted radiations. To meet this end, the scattered beams must propagate in a specific direction, particularly at an exit angle θ with respect to the diffracted plane of atoms that is equivalent to the angle θ of incident X-ray, wherein the scattered waves are thoroughly in phase to allow constructive interference.

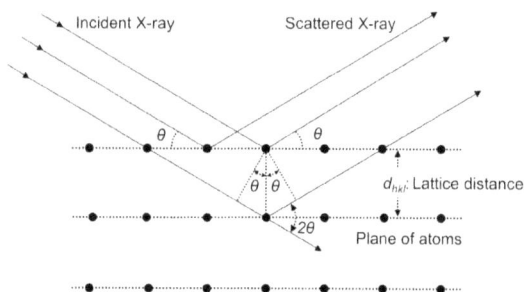

Figure 3. 2. Diffraction of X-rays by the lattice atomic planes in the crystalline powder specimen (*adapted from reference [105]*).

Significantly, such scenario depicted in Fig. 3.2 is formulated mathematically as

$$n_d \lambda = 2 d_{hkl} sin\theta_B \tag{3.3}$$

that is named *Bragg's law* after W. L. Bragg who first defined this relation. In this expression, λ is the wavelength of X-ray, θ_B the scattering angle (Bragg angle) and n_d is an integer representing the order of diffraction. On such basis, the collection of these diffracted X-rays in the measurement evidently allows one deriving the distribution of atoms in the materials.

Particularly, the powder method of X-ray diffraction is likely the most leading technique to this end. Literally, "powder" indicates that the monocrystalline solid under investigation is physically in the powder form, whereas such single crystals are hardly encountered in materials research. In a broader sense, this term refers to crystalline domains (grains) that are randomly oriented in a polycrystalline sample. In this sense, the method is likewise frequently employed for studying polycrystals in a variety of states, including bulk, thin films and particles suspended in a liquid.

Experimentally, X-rays diffracted by a powder smaple are registered by an electronic detector (counter in Fig. 3.1) implanted in a modern, computer-integrated diffractometer and expressed in a pattern plotted as the radiation intensity as a function of the diffraction angle (2θ). Structural analysis begins with fingerprinting the measured diffraction profile with the aid of a collection comprising an army of well-characterized diffraction patterns of diverse standards, which are summarized by the International Center for Diffraction Data (ICDD).[108] Particularly, the full agreement in the angular position and amplitude distribution of the collected diffraction lines with those in the diffraction pattern of a known substance substantiate the presence in the sample. Moreover, such correspondence allows resolving the lattice planes (hkl) and the associated interplanar distances (d_{hkl}) indexed respectively by distinct diffraction lines to specify the atomic arrangement. In consequence, the crystalline phase and lattice parameters of interest are accordingly concluded.

In contrast, additional important information regarding the grain size in polycrystalline materials is gleaned from the discrepancy in the line width of experimental Bragg diffractions against those in the benchmark diffractogram. Such angular divergence originates mostly from the attenuation (destructive interference) of scattered X-rays at angles deviated from the exact Bragg condition (Eqn. 3.3) more or less influenced by the periodicity of crystallographic planes within the grain.[102,105,106] This dependency renders the estimation of the crystal size (t_c) from the measured width (B) feasible using the well-known Scherrer formula.

$$t_c = \frac{0.9\lambda}{B\cos\theta_B} \tag{3.4}$$

In this equation, t_c represents the diameter of crystallites defined in a direction perpendicular to a set of Bragg planes that correspond to a specific diffraction line, B the full-width at half maximum

(FWHM) of this broadened diffraction peak, θ_B the associated Bragg angle, and λ is the wavelength of X-rays.

In this study, X-ray powder diffraction carried out on a STADI-P diffractometer (STOE & CIE, Darmstadt, Germany) using a Ge-monochromatized Cu-$K_{\alpha 1}$ radiation (40 kV, 40 mA) employed for probing the crystal structure and phase of the topical β-SnWO$_4$ spikecubes and Au@Nb@H$_x$K$_{1-x}$NbO$_3$ nanopeapods. Samples were prepared by depositing the powdery solids on an adhesive Scotch tape and immobilized via the attachment to a cellulose acetate flake. The diffractograms were registered within an angular interval of $5° \leq 2\theta \leq 70°$ and analyzed with the Win-XPOW software package via computerized searching the ICDD database for rational candidates.

3.2 Ultraviolet-Visible (UV-VIS) Spectroscopy

Ultraviolet (UV) and visible (VIS) radiation occupy the spectral regions distinct from that of X-rays at longer wavelengths in the electromagnetic spectrum with characteristic values of 10 to 400 nm for UV and 400 up to nearly 800 nm for VIS light, respectively.[101,109,110] Particularly, the regime from approximately 190 to 900 nm is employed in a commercial UV-Vis spectrometer as the working range in view of the absorption of UV with wavelengths less than 180 nm by the atmospheric gases.[110] Given these UV and VIS photons bearing respective energies up to 6.9 and 3.1 eV that are enough to promote electronic transitions in molecules and materials, UV-VIS spectroscopy is another ubiquitous technique for qualitatively characterizing the optical and electronic properties of materials.[109-113] Moreover, the linear relationship between the absorption

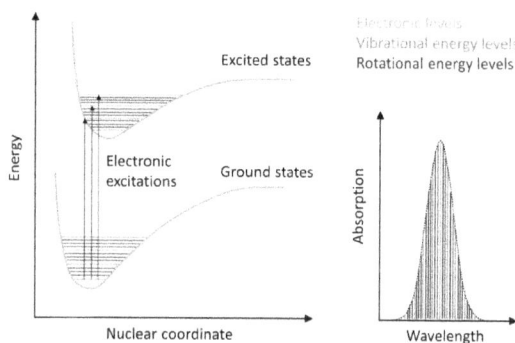

Figure 3. 3. The dispersive absorption band arises from the concurrent rotational, vibrational and electronic (ro-vibronic) transitions in molecules (*adapted from reference [111]*).

intensity and the absorber concentration (elucidated in the third paragraph) renders such method attractive for the quantitative analysis as well.

Given the electronic transition having characteristic energy difference specific to the material, the UV-VIS absorption spectrum can be useful for identification. This is true for atoms, lanthanide ions and some molecular complexes including transition metals, giving rise to sharp absorption bands at characteristic wavelengths in the absorption spectrum. However, for most molecular compound, electronic levels are superimposed by the vibrational and rotational energy levels, resulting in multiple transitions with diverse energies (Fig. 3.3). Consequently, the associated UV-VIS spectra mostly demonstrate broad features in absorption bands, leading to the necessity of coupling additional analytics (e.g. elemental analysis, Raman spectroscopy, nuclear magnetic resonance spectroscopy, etc.) for accurate qualitative description. Nevertheless, such characteristics otherwise facilitate the quantitative investigation. In this regard, UV-VIS spectroscopy is preferentially employed as a diagnostic tool to register the concentration variation of reactants or products during the reaction. In this application, absorption measurements are in general carried out at a single wavelength.

When the radiation interacts with the analytes, a number of processes occur including absorption, scattering, reflection, interference, fluorescence/phosphorescence (absorption and reemission) and photochemical reaction (absorption and bond breaking), leading to the decrement in light intensity. Particularly, only the loss of light owing to the absorption in order to prompt the charge transition is of particular interest.

Experimentally, UV-VIS spectroscopy measures such attenuation via the detector positioned downstream the sample and expresses the amount in terms of either transmittance (T_{ph}) or absorbance (A_{ph}).

$$T_{ph} = I_t / I_0 \qquad (3.5)$$

$$A_{ph} = -\log T_{ph} \qquad (3.6)$$

In Eqn. 3.5, I_0 and I_t are the incident and transmitted radiant energy registered on unit area in unit time, respectively. Evidently, the dissipation of light depends greatly on the extent of light-absorber interactions that is determined by the travel distance of the light through the sample (the interaction path length) and the concentration of the absorbing species in the matrix. Particularly, Lambert is credited for mathematically formulating the former effect for the first time.[111]

$$T_{ph} = e^{-kb} \tag{3.7}$$

In this expression, e is the base of natural logarithms, k a constant and b is the path length (in units of centimetres). The formulation describing the latter effect is otherwise put forward by Beer and the combination of these two laws gives the well-known Beer-Lambert law.

$$T_{ph} = e^{-\varepsilon_{molar} bc} \tag{3.8}$$

$$A_{ph} = -\log T_{ph} = \varepsilon_{molar} bc \tag{3.9}$$

In Eqn. 3.9, ε_{molar} is the wavelength-dependent molar absorptivity or molar extinction coefficient and expressed in units of L mol^{-1} cm^{-1}.[113] Significantly, this linear expression is the rule of thumb for the quantitative analysis that is primarily carried out on molecules and inorganic complexes in the liquid matrix.

Although quantification is of greater interest, clues to the chromophores (functional groups containing unsaturated π bonds) and the extent of electronic delocalization (conjugation) of the molecular system are alternatively available from the UV-Vis spectra.[111] Particularly, such qualitative study is otherwise favored in applied sciences (e.g. optoelectronics and solar energy conversions) on materials including plasmonic metals, semiconductors and so forth.[110] Such preference originates mostly from the distinct spectral features, e.g. narrow absorption bands localized at specific wavelengths and evident absorption edges, allowing for reliable qualitative description. In this sense, the spectral profile in lieu of absorbance grabs more attention to this end.

Information including physical dimension and morphology, the chemical composition and the surface state of the plasmonic metal can be resolved from the wavelength of the characteristic absorption peaks. In contrast, the band gap of a semiconductor is determined from the wavelength at the point of inflection on the featured absorption edge. Particularly, the measurement of powder semiconductors is in general carried out in the reflectance geometry that is in contrast to the molecular and metallic analytes measured in the transmittance mode. In this way, the extent of light extinction is expressed in terms of relative reflectance (R_∞), which is ascribed mostly to concurrent scattering and absorption processes.

$$R_\infty = R_{sample} / R_{reference} \tag{3.10}$$

In this expression, R_{sample} and $R_{reference}$ are the reflectance of the sample and the reference standard, respectively. The inherent scattering and absorption emerge from R_{∞} via the conversion to the Kubelka-Munk function ($F(R_{\infty})$),

$$F(R_{\infty}) = \frac{K}{S} = \frac{(1-R_{\infty})^2}{2R_{\infty}} \tag{3.11}$$

wherein K and S are the coefficients of absorption and scattering, respectively.[114]

To extract the absorptive component (K) of real interest from $F(R_{\infty})$, the sample under investigation is experimentally diluted with a "white" standard (e.g. barium sulfate) having a known S at different wavelengths. In this way, $F(R_{\infty})$ is exclusively K-dependent on the premise that S of the diluted sample can be approximated by that of the diluent. Significantly, the overall treatment faithfully deconvolutes the absorption from R_{∞}, which is then employed in the Tauc law that formulates the excitonic absorption in semiconductors due to the interband transition.[115-117]

$$(h\nu K)^{1/n} = A(h\nu - E_g) \tag{3.12}$$

In this expression, h is the Planck constant, ν the frequency of incident UV and VIS photon, A a proportional constant and E_g is the band gap of the semiconductor.

Particularly, exponent n in Eqn. 3.12 can be either 2 for the transition that demonstrates the changes in energy and momentum simutaneously (indirect interband transition) and 0.5 for the excitation carried out at the absence of momentum change (direct interband transition). This value is specific to material under examination. This equation turns into the following format

$$(h\nu F(R_{\infty}))^{1/n} = A(h\nu - E_g) \tag{3.13}$$

in view of K experimentally expressed by $F(R_{\infty})$. On such basis, the linear dependency of $(h\nu F(R_{\infty}))^{1/n}$ on $h\nu$ manifested in the consequent Tauc plot is employed to yield E_g of the material via extrapolating the line to the abscissa with variable $h\nu$.

In this work, the UV-VIS spectra of β-SnWO$_4$ and Au@Nb@H$_x$K$_{1-x}$NbO$_3$ in powder form were registered by a Cary 100 spectrometer from VARIAN (Palo Alto, USA) in a spectral interval of 250-900 nm. The measurements were carried out in a reflectance geometry with an integrating sphere from LABSPHERE (North Sutton, USA). The collected spectra were further resolved for qualitatively characterizing E_g of the inherent semiconductors. Additionally, quantitative studies of the molecular dyes including zwitterionic rhodamine B (RhB, $C_{28}H_{31}ClN_2O_3$, 99%) and cationic methylene blue (MB, $C_{16}H_{18}N_3SCl$, 95%) over the lapse of 1.5- and 4-h photooxidation reactions were performed in a dual-beam transmittance mode. The measurements were collected with collimated VIS-light beams at monochromatic wavelengths of 554 and 665 nm, respectively. These analytes were dissolved in water and hold in the 10-mm-path-length cuvettes made of quartz glass of quality 6Q. The deuterium and the quartz halogen lamps were employed as the UV- and VIS-light sources, respectively.

3.3 X-ray Absorption Spectroscopy (XAS)

In the former introduction to XRD, the utility originates primarily from the close match between the wavelength of X-ray and typical interatomic distance.[102-106] Thus, the positions of the constituent atoms in the basic unit of a periodic structure can be derived from the informative interference patterns. Moreover, another fortuitous property of X-ray, particularly the photon energy that is of the order of the core-level binding energies for essentially every element in the periodic table, reinforces the diagnostic power of the X-ray technique.[110,118-128]

Such property opens a way for X-ray-matter interaction, viz. absorption in addition to scattering. Particularly, X-ray is absorbed by an atom (analyte) distributed in either the chemical elements or compounds (matrix). In analogy to the UV/VIS-light-absorber interactions, electronic transitions from particularly the tightly bound core levels to either the vacant valence orbitals or continuum states are available via this new channel (left panels in Fig. 3.4).

In this connection, standard descriptive nomenclature for such process depends exclusively on the core electron participating in the excitation (Fig. 3.5). For instance, K-shell or K-edge absorption refers to the transition of electrons from the inmost $1s$ orbital (principle quantum number (n) of 1). At the L-shell ($n = 2$), electronic excitations can be promoted either from the $2s$ orbital, giving rise to the L_1-edge absorption. Alternatively, L_2 and L_3 absorption edges are attributed to the electronic transitions from the $2p_{1/2}$ and $2p_{3/2}$ orbitals formed upon spin-orbit coupling, respectively.

The attenuation of X-rays due to absorption after penetrating the specimen with thickness t_s follows the Lambert's law (Eqn. 3.7).[110,119,118-127]

$$I_t = I_0 e^{-\mu(E)t_s}$$
(3.14)

In this expression, $\mu(E)$ is the X-ray absorption coefficient, I_0 and I_t are the X-ray intensities incident on and transmitted through the sample, respectively. Given the specific energy difference to the electronic transition, such energetic dependency of the probability for X-ray absorption manifests substantially in $\mu(E)$, rendering this variable of primary interest in the measurement. On

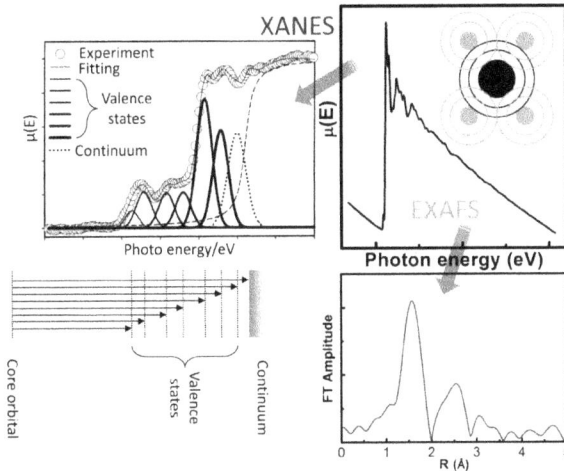

Figure 3. 4. (Upper right panel) X-ray absorption fine structure (XAFS) divided into (blue tinted region) the X-ray absorption near edge structure (XANES) and (yellow tinted region) the extended X-ray absorption fine structure (EXAFS) regions. (Upper left panel) Deconvoluted XANES demonstrates the overall X-ray absorption originated from (lower left panel) diverse electronic excitations. (Inset in upper right panel) Schematic illustration of the interference of outgoing (black circles) and backscattered (green circles) photoelectron waves leads to the oscillatory behavior of EXAFS. (Lower right panel) Fourier transformed EXAFS reveals the coordination shells around the absorbing atom (*adapted from reference [118-128]*).

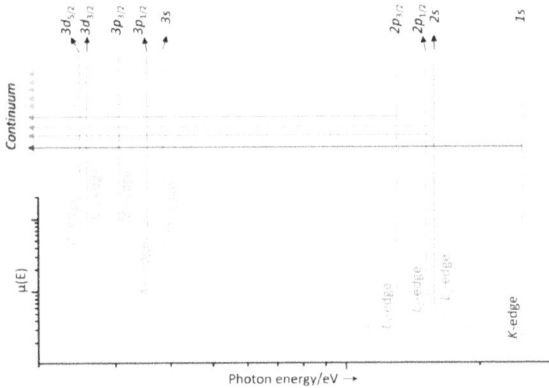

Figure 3. 5. Schematic diagram of saw-tooth-like X-ray absorption edges of an isolated atom at discrete photon energies corresponding to the characteristic binding threshold of particular atomic levels (*adapted from reference [120]*).

such basis, the number and distribution of available valence states can be derived from the modulation of $\mu(E)$ at discrete photon energies, which is of great interest to the coordination chemistry (left upper panel in Fig. 3.4).

Significantly, K and L_1 absorption edges are preferentially employed in this approach to estimate those states comprising p orbitals. By contrast, L_2 and L_3 absorption edges are otherwise utilized to evaluate the valence states bearing s or d orbitals. Such discrepancy originates mostly from the electronic transitions subjected to the well-known selection rules for electric dipole interactions. Particularly, the Laporte rule ($\Delta l = \pm 1$) dictates the K- and L_1-edge absorption (s states with orbital angular momentum quantum number (l) of 0) ascribed virtually to the transitions between shells $s \rightarrow p$ while the L_2- and L_3-edge absorption (p states with $l = 1$) is attributed substantially to the transitions between shells $p \rightarrow s$ or $p \rightarrow d$, respectively.

Noteworthily, ionization of the atomic level commences at photon energies equivalent to the binding threshold of the core electrons, giving rise to a tremendous increment of $\mu(E)$ due to the emission of photoelectrons toward the vast continuum states. This step-like spectral feature is named the absorption edge (Fig. 3.6a).

Above this edge, excess photon energy is converted into kinetic energy of the photoelectrons propagating away from the atom in the form of an outgoing spherical wave. When an additional atom is in the close proximity, the photoelectron wave is scattered by the electrons of this neighboring atom and returns back to the absorbing one (Fig. 3.6b). The interference of these photoelectron wavelets perturbs $\mu(E)$ at the absorbing site, provided that the presence of the photoelectron scattered back to the absorbing atom modulating the availability of the electronic state that is necessary to carry out photoionization. Given the absorption measurement in general carried out in a photon energy scanning method, the energy along with the corresponding wavelength of the photoelectron accordingly alter.[110,119,123-126] Consequently, the interference of the outgoing and backscattered photoelectron wavelets continuously goes in and out of phase, giving rise to the oscillatory behavior of $\mu(E)$.

Evidently, the interference pattern is of high relevance to the local arrangement of atoms that trigger the scattering. In other words, short-range structural information complementary to the long-range clues gleaned via the XRD technique can be extracted from the absorption measurement. Moreover, the backscattering amplitude and the phase shift as a function of the photon energy are highly element-specific. Such dependency further renders the elements making up the overall coordination environment around the absorbing atom distinguishable.

Although X-ray absorption occurs always due to an effective electronic transition, the collected full spectrum designated as X-ray absorption fine structure (XAFS) is in general divided into two regimes in order to derive diverse informations of interest.[110,118-128]

The X-ray absorption near edge structure (XANES) comprises literally the spectral features in close proximity to the main absorption edge over an energy span of ca. 30 eV before and beyond the edge. X-ray absorption within this region is characterized primarily by the electronic transitions to the empty valence states, rendering the empirical analysis of the electronic structure feasible. Moreover, these states are highly susceptible to the chemical bonding and the oxidation state of the atomic absorber, such sensitivity allows the evaluations of the formal valence state and the coordination context.

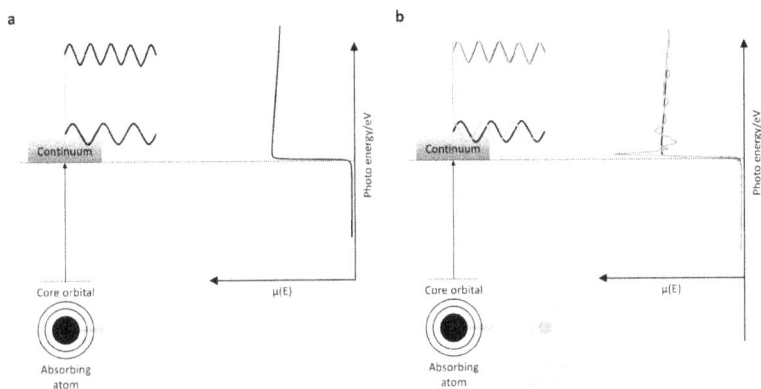

Figure 3. 6. (a) Schematic diagram of X-ray absorption via the photoelectric effect. The tightly bound core electron is liberated from the absorbing atom in this process, giving rise to the photoelectron. The photoelectron travels outward in the form of a spherical wave. (b) In this simplest picture, the outgoing photoelectron wave (black circles/wave) is scattered by an adjacent atom as a point obstacle and return back to the absorbing one (cyan circles/wave). The consequent interference modulates the amplitude of the photoelectron wave function and in turn $\mu(E)$ of the absorbing atom (*adapted from reference [119]*).

By contrast, the extended X-ray absorption fine structure (EXAFS) refers verbally to the oscillatory spectral features present at ca. 30 eV above the edge. At higher photon energies, X-ray absorption arises substantially from the photoelectric effect ejecting the photoelectron to the continuum states. EXAFS is thus irrespective of coordination chemistry but depends exclusively on the local atomic arrangement around the absorbing atom. Quantitative information including the coordination number, interatomic distances, structural and thermal disorders close to the absorbing site can then be extracted.

Evidently, the principal challenge to experimentally collect the XAFS lies in the radiation source to provide a broadband X-ray necessary for the energy scan. Such requirement is difficult to meet with conventional laboratory X-ray tube but satisfied by the employment of a synchrotron facility.[110,119,123-126]

In a modern synchrotron, polychromatic radiation is produced in this way: electrons emitted from the source are first accelerated in a linear accelerator. The velocity along with the energy is further promoted after entering a booster ring. Afterwards, those charged particles are transferred to the storage ring implanted with a series of bending magnets that force the electrons to circulate over a million times each second. Particularly, the electromagnetic radiation is emitted upon the change in the moving direction of the charged particles due to traversing the magnetic field. The synchrotron radiation is thus characterized by a continuous energy spectrum ranging from infrared to hard X-rays, high intensity, strong polarization and a pulsed nature.

The radiation is eventually delivered through the beamlines to a number of end-stations for usage in a variety of experiments. Hence, each beamline is particularly configured in order to meet the characteristics of the analytic technique. In a typical XAS beamline, mirrors are employed to collimate and focus the X-ray and the apertures and the slits are utilized to define the beam size. A double-crystal monochromator that uses *Bragg's law* (Eqn. 3.3) to diffract X-ray is implanted to allow energy selection through the desired scan range.[102-106] The XAFS of the specimen mounted on a sample stage in the end-station is collected either in a transmission mode, wherein the intensities of incoming and transmitted X-ray are measured to derive $\mu(E)$ as a function of photon energy via Eqn. 3.14.[110,119,123-126] Alternatively, the decay products in the X-ray absorption process including either fluorescent X-ray measured in a fluorescence mode or Auger electrons collected in an electron yield mode provide additional approaches to indirectly estimate $\mu(E)$.

In this work, the synchrotron radiation at the National Synchrotron Radiation Research Center (NSRRC) in Taiwan was employed as the energy-tunable X-ray source to collect the XAFS of β-SnWO$_4$ at the W L_3-edge and O K-edge, respectively. The W L_3-edge absorption was measured at beamline 17C in a total fluorescence yield (TFY) mode under ambient conditions.[129] The radiation from the storage ring was monochromatized by a Fixed-Exit Si(111) double-crystal monochromator and the intensity of X-ray was measured by two ionization chambers filled with helium gas. The measurement was carried out in a "normal incidence-grazing exist" geometry, in which solid β-SnWO$_4$ was illuminated by the X-ray at nearly normal incident angle and the detector was placed at a grazing exit to collect the fluorescence signal. Particularly, a filter with a slit of 5~10 mm was placed before the detector. In this way, the detection integral was refined upon circumventing the probable attenuation of $\mu(E)$ stemmed mostly from self-absorption effect.

The EXAFS region of the collected XAFS was subsequently processed and quantitatively analyzed using the Athena data analysis package implemented with the IFEFFIT codes.[130] In this way, the local coordination structure around the absorbing W element in the β-SnWO₄ matrix was derived.

By contrast, O K-edge absorption was measured at beamline 20A implanted with a horizontal focusing mirror (HFM), a vertical focusing mirror (VFM), a spherical grating monochromator with 4 gratings, and a toroidal refocusing mirror to ensure a mean energy resolution of 5000.[131] XAFS was collected in a total electron yield (TEY) mode at room temperature. A geometry of powdery β-SnWO₄ fixed by the Scotch tape on the sample holder in an ultrahigh vacuum (UHV) chamber was employed to carry out the measurement.

The XANES region of the collected XAFS was numerically analyzed via the real space full multiple scattering (FMS) simulation.[132,133] In this way, the projected density of states (pDOS) of O $2p$, W $5d$ and Sn $5s/5p$ orbitals making up the unoccupied states of the β-SnWO₄ matrix was deconvoluted.

Likewise, the XAFS of Au@Nb@H$_x$K$_{1-x}$NbO₃ was collected at the NSRRC in Taiwan but at beamlines 16A and 01C at the Nb L_3-edge and Au L_3-edge, respectively. Particularly, the measurements were carried out in the presence and absence of additional simulated sunlight.[134,135] The XANES region of the collected XAFS was employed for comparison to reflect the availability of electronic states at the Nb $4d$ and Au $6s$-$6p$-$5d$ hybridized orbitals, respectively.

3.4 Dynamic Light Scattering (DLS)

The static light-scattering in classical text, particularly the well-known Rayleigh scattering, underlies the dynamic light scattering (DLS) method employed for the size distribution analysis of particles in solution or suspension.[112,136-138] Rayleigh's theory dated back to 1871 deals mostly with the scattering from the particles that are much smaller compared to the wavelength of the light. In such a case, the scattered beams propagate in all directions with the perturbations in light intensity (I_s).

$$I_s = I_0 \frac{1+\cos^2 \theta}{2R_p^{\ 2}} (\frac{2\pi}{\lambda})^4 (\frac{n^2-1}{n^2+2})^2 r^6 \tag{3.15}$$

In this expression, I_0 and λ is the intensity and wavelength of the incident light, R_p the distance to the particle, θ the scattering angle, n the refractive index of the particle and r is the radius of the particle.

In practice, such colloidal sized particles are no longer stationary when distributed in solution but undergo random motions due to the multiple collisions with the thermally driven solvent molecules, the well-known Brownian motion.[112,137,138]

$$(\overline{\Delta x})^2 = 2\Delta t \tag{3.16}$$

In Eqn. 3.16, $(\overline{\Delta x})^2$ is the mean squared displacement in time t, and D is the diffusion coefficient. Particularly, D is of relevance to the hydrodynamic diameter d of the particle,

$$D = \frac{k_B T}{3\pi\eta d} \tag{3.17}$$

wherein k_B is the Boltzmann constant, T the temperature and η is the viscosity of the solution.

Consequently, the particles constantly move, giving rise to the Doppler shift between the wavelengths of the incident and scattered light, respectively. Moreover, such temporal fluctuations are likewise present in light intensity in terms of the interference of the light scattered from respective colloids continually changed with the variable interparticulate distance. The rate of such intensity modulation depends on the velocity of the Brownian motion, wherein smaller particles diffuse faster than larger ones (Eqn. 3.16-17). On such basis, the size distribution of the particles in solution can be extracted from this time-dependent perturbation of the scattered light.

Experimentally, the specimen under investigation is exposed to the monochromatic light emitted from a laser with the wavelengths of either 633 or 532 nm.[137-140] The light scattered by the analytes follows the Rayleigh scattering but only the radiation exiting at a particular scattering angle θ with respect to the incident beam is measured by the detector. The scattering intensity is collected at consecutive time intervals in order to derive the rate of the intensity fluctuation via a digital processing board. Such information is eventually utilized by a software package to determine the particle size. Three distribution indexes are in general employed in the particle size analysis including the intensity, the volume and the number distributions, respectively.

Literally, the intensity distribution refers to the size distribution weighted by the scattering intensity, leading to a six-power dependence of such index on the particle size according to the Rayleigh approximation (Eqn. 3.15). Likewise, the volume distribution verbally suggests the size distribution weighted by the volume of the particle, giving rise to a three-power dependence of this index on the particle size. In contrast, the size weighting is absent from the number distribution, resulting in this index faithfully reflecting the absolute number of the particles.

The DLS method employed in the β-SnWO$_4$ and Au@Nb@H$_x$K$_{1-x}$NbO$_3$ studies for the size analysis was carried out using the Nanosizer ZS apparatus from Malvern Instruments (Herrenberg, Germany). This equipment was implanted with a 4.0 mW He-Ne laser emitting red light at the wavelength of 633 nm. The scattered light through those topical artifacts in the suspension form hold in the cuvettes was collected by a detector positioned at a scattering angle of 173°. The number weighted distribution is utilized in the overall particle size analyses.

3.5 Scanning Electron Microscopy (SEM)

In the former introductions to diverse analytic techniques, the electromagnetic radiations typified by the wavelength (λ) including X-ray and UV-VIS light were employed as the exclusive probes. Although an electron is in general regarded as a particle carrying a single negative charge (e) of 1.6 $\times 10^{-19}$ coulomb (C) and having a rest mass (m_e) of ca. 9 $\times 10^{-31}$ kilogram (kg), such wave nature is likewise present in this elementary particle.[102,110,141,142] Such wave-particle duality renders the electron technique absolutely powerful and flexible for material characterization, as elaborated below.

The electron employed in the scanning electron microscopy (SEM) has a characteristic kinetic energy of 0.1 to 30 keV that arises from traversing an electric field with an accelerating voltage (V). Particularly, the wavelength (λ) of the electron is dictated by the momentum (p) according to the *de Broglie* formula.

$$\lambda = \frac{h}{p} = \frac{h}{mv} \tag{3.18}$$

In this expression, h is the Plank constant, m and v are the mass and the velocity of the electron, respectively. Given the most useful accelerating voltage (typically few tens of kV) for SEM resulting in v a significant fraction of the velocity of light (c), the relativistic effect leads to the increment in m.

$$m = \frac{m_e}{[1-(v/c)^2]^{1/2}} \tag{3.19}$$

Given the kinetic energy of the electron otherwise associated with the relativistic change of mass,

$$qV = (m - m_e)c^2 \tag{3.20}$$

the exclusive dependence of λ on V is readily worked out upon the combination of Eqn. 3.18-20.

$$\lambda^2 = h^2 / (2qVm_e + q^2V^2/c^2) \tag{3.21}$$

$$\lambda = [1.5/(V + 10^{-6}V^2)]^{1/2} \tag{3.22}$$

Eqn. 3.22 is derived from the substitution of the values for h (6.62 x 10^{-39} Js), q (1.602 x 10^{-19} C), m_e (9.108 x 10^{-31} kg) and c (2.998 x 10^8 ms^{-1}) in Eqn. 3.21, respectively.

Microscopy literally refers to the technical field of employing the microscope to transform the objects of interest that are beyond the resolution limits (ca. 0.1 millimetre at a viewing distance of 25 centimetre) of the naked eye into an image. The resolution limit is defined as the minimum distance (d_{min}) by which two objects can be separated and still appear as two distinct entities. A microscope is an optical system designed for promoting the resolving power that is formulated mathematically as

$$d_{min} = \frac{0.612\lambda}{n \sin \alpha_{con}} \tag{3.23}$$

via the employment of the lens. In this expression, λ is the wavelength of the imaging radiation, n the relative refraction index of the medium between the object and the lens, and α_{con} is the semi-angle of the cone of the radiation from the object plane accepted by the lens.

Evidently, the resolution limit reduces with the decrement in λ and the increments in n and α_{con}. In this sense, the resolving power of an electron microscope far exceeds that of an optical microscope (ca. 150 nm using the green light source with the wavelength of 400 nm), provided that the derived wavelength of the electron from the characteristic working voltage in SEM (Eqn. 3.22) is of the order of few thousandths to few tenths nanometres that is significantly small in comparison with that of the visible light employed in an optical microscope. Particularly, SEM takes advantage of such exceptional resolution for the material investigations on the processing, the properties and the behavior that depend remarkably on the microstructure.

The experimental configuration of a modern scanning electron microscope (SEM) is depicted in Fig. 3.7. On the top of the column is the electron gun that produces and endows the electrons with the kinetic energy of 0.1 up to 30 keV via an external bias. Several electron sources including thermionic tungsten (W) and lanthanum hexaboride (LaB$_6$) electron guns are available to SEM while the field emission gun is the mainstream in the modern SEM. The electromagnetic lenses and

38

the apertures underneath the electron gun are aimed at focusing the diameter of the electron beam to the desired and, eventually, smallest possible size. The electrons are deflected due to the magnetic field produced upon applying a current to those electromagnets made of the coils of wire.

In consequence, the electron beam is focused to a spot, wherein the size (1-100 nm in diameter) refers to the probing area on the materials. Significantly, the smaller the probing area is, the superior the resolution and the magnification can be. Moreover, the small convergence angle (α_{con}) of the electron probe suggests an insignificant change of the spot size along the beam direction. In other words, the resolution is yet agreeable when the sample lies more or less above or below the focal plane. The range of such deviation is termed the depth of field (h_{df}).

$$h_{df} = \frac{d_{min}}{\tan \alpha_{conv.}} \qquad (3.24)$$

Such advantage alternatively favors the three-dimensional information extracted from the plane image. A high column vacuum of 10^{-6}-10^{-7} Torr is indispensable for the electron beam to avoid defocusing of the electron beam by scattering of the electron by the gas molecules. The electrons impinge on the bulk specimen and interact with the sample in the so-called interaction volume. Its size depends on the material composition, the energy and the incident angle of the

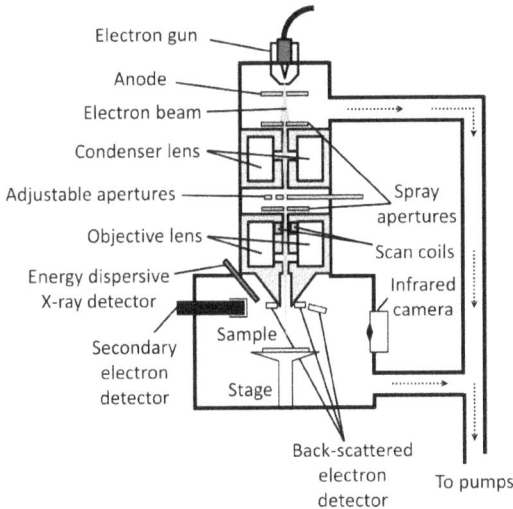

Figure 3. 7. Schematic overview of a scanning electron microscope (*adapted from reference [102,110,141,142]*).

the electron beam. However, only the electrons successfully escaped from the sample via either elastic or inelastic scattering carry useful information for analysis. Such emission volume is in general a fraction of the total size of the interaction volume that depends strongly on the scattering mechanism (Fig. 3.8).

Backscattered electrons (BSEs) have, by definition, a kinetic energy more than 50 eV. A special case is the elastically backscattered electrons, which carry information of the average atomic number and material density of the specimen. Secondary electrons (SEs) are otherwise emitted away from regions close to the sample surface due to their small kinetic energy. In contrast, high-energy BSEs may exit from the sample even if they are generated at a large distance from the sample surface. On such basis, low-energy SEs are mostly employed to reveal the topographical features including the surface texture and roughness of the materials. By contrast, high-energy BSEs are preferentially utilized to disclose the chemical compositions of the specimen. Noteworthily, this information discussed herein originates only from one single spot that the electron beam impinges. To form an image, the electron probe needs to move from place to place across the specimen. Particularly, such scanning process is carried out linearly via the magnetic deflection by the scan coils (Fig. 3.7) over a rectangular domain.

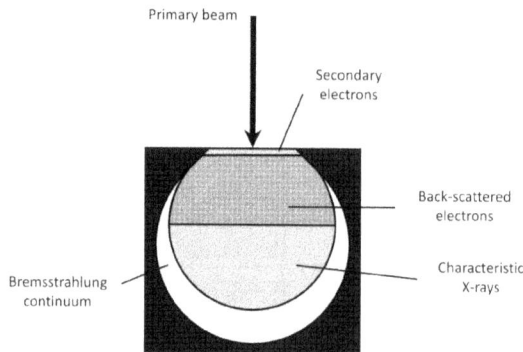

Figure 3. 8. Electrons and X-ray photons are generated upon the electron-probe-specimen interaction. The characteristic tear drop interaction volume depends on their physical properties and energies (*adapted from reference [102,110,141,142]*).

Detectors are used for the collection of the electrons emitted from the sample. This includes the Everhart-Thornley (ET) detector and the annular solid-state BSE detector, respectively. The ET detector for the SE collection is classically arranged below the lenses and apertures in an asymmetric geometry. Particularly, the ET detector highlights the material topography (e.g. edges, corners, steps and surface roughness) and the collection efficiency is strongly dependent on the relative position of those physical features and the detector (Fig. 3.9). Moreover, modern SEMs contain in addition a so-called "in-lens" detector. The electrons are efficiently collected by the field of the objective lens that leads to an image with an improved signal-to-noise ratio.

The SEM image is built upon the registered intensities of those informative electrons, which is eventually projected onto the viewing cathode ray tube (CRT). Particularly, the photograph is constructed in a scanning method that mimics the movement of the electron beam across the specimen. Moreover, bilateral scanning procedures synchronize, leading to a point-to-point transfer of the intensity information to the final picture. On such basis, the material under investigation is manifested in a grayscale image, wherein the magnification is defined as the side length of the photograph divided by that of the scanned region on the specimen. Given the invariable dimension of the recording device, magnification is modulated via readily altering the size of the scanned region. A typical magnification range for the SEM is ten- to sub-million-fold.

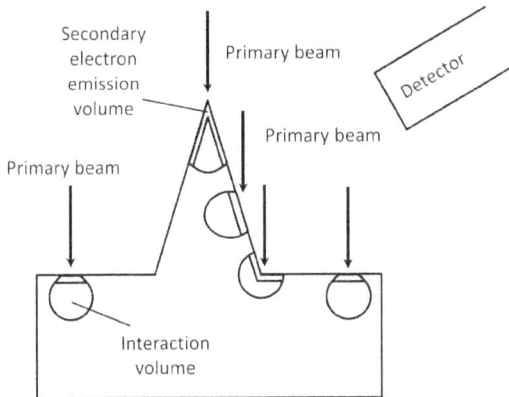

Figure 3. 9. The dependency of the secondary electron collection efficiency on the specimen-detector geometry, wherein most electrons emitted from the leftest tear drop interaction volume are prevented from the detector by the tip. More significantly, the dependency of the secondary electron yield on the incident angle of the primary electrons, wherein more secondary electrons are available from shape tip, tilted surface and the corner than that from a flat plane (*adapted from reference [142]*).

The contrast level of a grayscale SEM photograph is dictated by the number, the emission angle, and the energy of the signal electrons. Significantly, the interpretation of the contrast depends strongly on the distinct behaviors of the SE and BSE, resulting in a variety of highly informative images. For instance, the number factor literally refers to the quantity of the electron sampled from the material, which underlies the contrast in the BSE compositional image in terms of the yield highly element-dependent. By contrast, the emission angle factor is of great relevance to the probe-specimen-detector configuration, provided that only a fraction of ejected electrons are available to the detector due to the limited size and shape. On such basis, the asymmetric geometry particularly employed by the standard ET detector is aimed at punctuating this factor, leading to the minor irregularities on the material surface manifested in high contrast in the SE topographic image (Fig. 3.9).

In this monograph, the topography of the topical β-SnWO$_4$ was registered in the SE detection mode using a scanning electron microscope (Supra 40 VP) from Zeiss (Oberkochen, Germany). Given the characteristically small sampling volume of the SE effectively quenching the overlap of information from closely neighbored surface features (Fig. 3.8), the morphology of β-SnWO$_4$ was successfully resolved in detail in the final photograph. A Schottky field-emission cathode was implemented in this SEM as the electron gun with the working voltage spanned from 1 up to 20 kV. The ET and the in-lens detectors were employed for the SE collection. The measurement was performed on the powdery β-SnWO$_4$ immobilized on a smooth silicon wafer fragment that was attached to the aluminum sample carrier from PLANO (Wetzlar, Germany) via conductive silver adhesive. The fixation was carried out via drop-casting the β-SnWO$_4$ suspension and the natural evaporation of ethanol solvent following up.

3.6 Energy Dispersive X-ray Spectroscopy (EDXS)

In the foregoing section, the emphasis was placed exclusively on the SEM image built upon the informative electrons that undergo either elastic or inelastic scattering in the material. Particularly, the energy loss of the primary beam in an inelastic process can be transferred to the core electron of the specimen atom to allow the ejection from the inner shell.[102,110,141] The created core hole is then refilled by the outer shell electron at a higher energy level, which in turn relaxes the excited atom. Excess energy is given out in this relaxation in the form of emitting either a photon (the characteristic X-ray) with energy equivalent to the difference in binding energy of the two electron levels that participate in the transition. Alternatively, an electron (the Auger electron) can likewise be ejected from the atom with the kinetic energy equal to that of aforementioned X-ray minus the binding energy of the shell, wherein the Auger electron initially resides.

Given the X-ray photons and the Auger electrons having energy specific to the atom, they are preferentially employed for identification of the elements within the material matrix.[110,141] Noteworthily, all the elements on the periodic table except hydrogen (H) and helium (He) have at least one characteristic X-ray line with the photon energy less than 10 keV. This phenomenon absolutely favors such chemical analysis carried out by the SEM with available working voltage up to 30 keV, provided that the emission efficiency reaches the maximum at the specimen bombarded by the electron beam with the kinetic energy nearly three times that of the emitted X-ray photon. Such preference leads to the majority of commercially available SEM nowadays implemented with an energy dispersive X-ray detector to further reinforce the diagnostic power of the electron microscopes.

The working principle for the X-ray detection is built upon each incoming photon producing a number of electron-holes pairs in the semiconducting detector that consists of either silicon (Si) or germanium (Ge). Particularly, the number of the photogenerated charges depends on the energy of the detected X-ray photon (in silicon such excitation takes 3.8 eV for one electron-hole pair formation). A bias on the semiconductor separates the electrons and holes, giving rise to the charge flow with the magnitude faithfully reflecting the X-ray photon energy. Given the low resistivity of Si, the photogenerated current is significantly diluted by the charge flow triggered merely due to the presence of the bias. Such noise can be experimentally minimized upon cooling the overall detector to liquid nitrogen temperature (77 K). To prevent the condensation of likely contaminants on the very cold surface of the detector, a thin protection window made of either beryllium (Be) or a polymer is additionally placed before the detector. Such design somewhat compromises the detectability to the light elements with characteristic Z less than 10 in terms of the significant absorption of Be and carbon (C) toward low energy X-rays.

Noteworthily, the registered intensities of diverse X-ray photons are in general translated into a set of peaks dispersed over an energy spectrum, wherein the energetic distribution is employed particularly in a qualitative analysis to fingerprint the elements present in the material. Alternatively, the intensity information can likewise be translated into the bright dot superimposed on a SE topographic image, wherein the dot density is a qualitative measure of the concentration of the constituent element. Such technique is termed X-ray mapping that aims particularly at highlighting the respective distribution of individual element over the material on a grey level or color scale. Significantly, the spatial resolution of the X-ray map is limited to 1 μm, provided that a high energy electron probe is essential to trigger efficient emission of characteristic X-ray, which in turn leads to the smallest sampling volume of approximately 1 μm^3 for a practical analysis.

In addition to the foregoing qualitative analyses, elemental quantification can likewise be carried out with the registered X-ray intensities otherwise. Particularly, quantitative analysis calls for a great deal of care in the experimental conditions in order to perform a nearly artefacts-free measurement, as elaborated below. In general, the EDXS detector is implemented in the SEM in an asymmetric geometry as the standard ET detector, which is close to the sample (Fig. 3.7). Such configuration provides an intuitive overlook of the specimen and moreover, favors the X-ray collection efficiency. Nevertheless, this arrangement otherwise accentuates the emission angle factor introduced in the preceding discussion on the contrast in the SEM image. In other words, X-ray emitted from the regions on the specimen, which are not in the line of sight, is unavailable to the EDXS detector. Such effect turns out to be significant on a rough sample, highly likely giving rise to the topographic artifact in the quantitative analysis.

This issue can be experimentally ruled out upon performing the measurement preferentially on the specimen this is flat on the scale of the electron beam diameter. Moreover, a flat sample for analysis can concurrently circumvent the radiated X-ray passing through the specimen again. In this way, the X-ray fluorescence referring to secondary characteristic X-ray originated from the interaction between primary X-ray and the specimen atom is effectively quenched.

The quantitative study on a flat sample is in general performed in this way: characteristic X-rays from the sample that is mostly a chemical compound are first measured for a finite duration. Prolonged detection allows the intensity registered in the energy spectrum becoming much pronounced and readily discernable. The concentration is then derived from the comparison with that collected under identical condition from a standard compound with a well-defined stoichiometry, which is known as the Cliff-Lorimer ratio method. Significantly, the discrepancies in the density and the average atomic weight between the material under investigation and the standard compound call for the corrections to the comparison, which is known as the ZAF technique.

The first letter, Z, refers to the atomic number (Z) correction to the X-ray generation efficiency of an atom, provided that the local atomic weight of the matrix dictates the extent of the primary electrons available for the X-ray generation and in turn the sampling depth (Fig. 3.8). The second character, A, refers to the absorption correction to the X-ray emission efficiency due exclusively to the X-ray absorption by the matrix. Particularly, the magnitude of A turns out to be significant for the analysis of a light element present in a heavy atom matrix in view of the heavy element readily absorbing low energy X-rays emitted from the lighter one. The last letter, F, refers to the secondary X-ray fluorescence that is a minor factor in comparison with A, provided that fluorescence is a very inefficient process. Noteworthily, such correction procedure is nowadays computerized in the modern SEM. More importantly, the employment of the "virtual standards" allows the foregoing comparison process likewise computerizable, leading to the quantitative analysis remarkably simplified and highly reliable.

In this dissertation, characteristic X-rays emitted from the heavy elements in the topical β-SnWO$_4$ artifact were employed for describing the chemical composition. The EDXS detector from AMETEC (Berwyn, USA) mounted on the Supra 40 VP SEM was utilized to this end. In this quantitative analysis, the working voltage and distance were 20-30 keV and ca. 8.5 mm, respectively. The reliability of the analysis was otherwise ensured upon pressing powdery β-SnWO$_4$ solid into a smooth pellet that was then immobilized on the conductive carbon tape from PLANO (Wetzlar, Germany) and attached to the aluminum sample carrier.

3.7 Transmission Electron Microscopy (TEM)

In analogy to scanning electron microscopy (SEM) introduced in the foregoing section, material characterization by virtue of transmission electron microscopy (TEM) likewise employs a beam of electrons directed at the specimens under investigation. Nevertheless, the electrons employed in TEM are in general accelerated to an energy up to the order of 100 to 300 keV.[102,110,141] In consequence, the penetration capability of such high-energy electrons renders TEM virtually a bulk technique, giving access to micro- or ultrastructural details internal to materials having a thickness no more than a fraction of a micron.[110,141]

The electron beam passing through such thin specimen contains two components, elastically and inelastically scattered electrons, respectively. Given the wave nature of the electron, the argument used to explain the diffraction of X-rays is likewise applied to interpret the electron-matter interactions. Concretely, any scattered electron waves that are in phase with one another will reinforce, leading to a strong beam of electrons. Contrarily, the scattered waves that are out of phase will not reinforce. The condition for reinforcement follows *Bragg's law* (Eqn. 3.3), suggesting that elastically scattered electrons at the Bragg angle (θ_B) most likely emerge from the specimen. Given the very short wavelength of the electrons employed in TEM (Eqn. 3.22),[141] Eqn. 3.3 is thereof further simplified and written as

$$\lambda = 2d_{hkl}\theta_B \qquad (3.25)$$

In this equation, λ and d_{hkl} are the wavelength of the electron and the interplanar distance of the crystal lattice (*hkl*), respectively. Particularly, Eqn. 3.25 suggests only the planes of atoms that are nearly parallel to the electron beam giving rise to strong diffraction.

The analysis of the spatial distribution of these scattered electrons, which is also known as an electron diffraction pattern, reveals a great deal of information about the atomic arrangement in the specimen, as elaborated below. These diffracted electrons through an angle of $2\theta_B$ by the crystal planes of spacing d_{hkl} hit the photographic negative in the TEM, which positions at a distance L from the specimen, at A (Fig. 3.10). By contrast, incident primary electron pass through the specimen without interaction hit the screen otherwise at O. The distance from the diffracted to this undiffracted beam on the electron diffraction pattern is r_{ED}. Simple geometry suggests the diffraction angle ($2\theta_B$)

$$\frac{r_{ED}}{L} = 2\theta_B \qquad (3.26)$$

The combination of Eqn. 3.25 and Eqn. 3.26 gives

$$r_{ED}\,d_{hkl} = \lambda L \qquad (3.27)$$

Eqn. 3.27 is referred to as the camera equation and suggests that r_{ED} is inversely proportional to d_{hkl}. More importantly, it allows one to determine d_{hkl} via measuring r_{ED} on the electron diffraction pattern in the TEM. The approximate value ($\pm5\%$) of λL, which is known as the camera constant (in units of nano- to centimetres), is in general synchronously displayed on the console of a modern TEM during the measurement.[110]

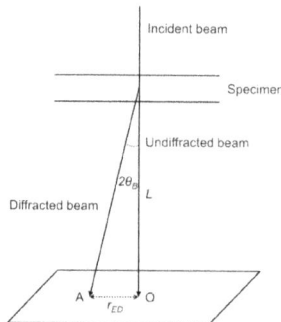

Figure 3. 10. Geometry for the formation of the electron diffraction pattern (*adapted from reference [141]*).

A well oriented monocrystalline specimen with several sets of crystal planes that are approximately parallel to the electron beam gives rise to a diffraction pattern consisting of a regular array of diffraction spots (Fig. 3.11a). Significantly, the vector pointing from the reflection of the undiffracted electrons to the corresponding Bragg reflection is oriented in the direction perpendicular to the crystal planes with a distance of $\lambda L/d_{hkl}$ (Fig. 3.11b). This is readily reminiscent of the reciprocal lattice that employs a point, which locates at a distance of $1/d_{hkl}$ from the origin, to represent the real crystal plane (Fig. 3.11c). The diffraction vector describing the position of such

46

reciprocal lattice point with respect to the origin is in general denoted by **g**. Such strong resemblance renders a straightforward interpretation of the electron diffraction pattern in terms of the reciprocal lattice. Labeling individual diffraction spots with appropriate crystal planes (*hkl*) begins with identifying the transmitted beam that is in general the brightest spot in the center of the diffraction pattern. In the closest proximity to this spot, two independent diffraction points are subsequently indexed in terms of **g**. The indices of other diffraction spots are eventually derived via linear combination of these two diffraction vectors. Most of the work in this procedure involves measuring the angles and the distances between the reciprocal lattice vectors. The identification is completed at specifying the normal to the plane of the electron diffraction pattern, which is known as the zone axis.

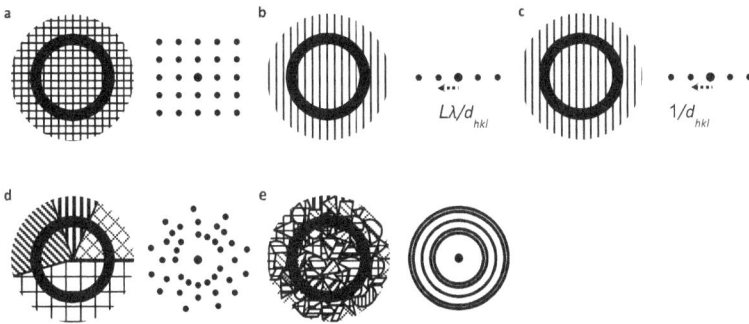

Figure 3. 11. (a) Diffraction pattern (right panel) of a single perfect crystal specimen (left panel), (b) wherein the vector pointing from the reflection of the undiffracted electrons to the corresponding Bragg reflection (right panel) is oriented perpendicularly to the lattice plane (left panel) with an interval of $\lambda L/d_{hkl}$. (c) This set of crystal planes (left panel) are otherwise expressed in terms of a series of reciprocal lattice points (right panel) that likewise aligned in an orthogonal direction with a spacing of $1/d_{hkl}$. (d) Diffraction pattern (right panel) of a specimen containing a small number of crystallites (left panel). (e) Those diffraction points start to merge into rings when a large number (≥ 5) of randomly oriented grains are present in the specimen (*adapted from reference [141]*).

Such index practice becomes complicated when the crystalline specimen contains a small number of grains that are diversely orientated, provided that the electron diffractogram in such a case is the sum of individual diffraction patterns (Fig. 3.11d). Significantly, the spots are not arbitrarily distributed but fall on the ring with a constant distance to the undiffracted spot. Moreover, those points start to merge into rings when a large number of randomly oriented grains are present in the specimen (Fig. 3.11e). In these cases, a computer program to calculate such diffraction patterns significantly simplifies the analysis.

Although quantitative diffraction information is more readily derived from X-ray than electron techniques, electrons have an important advantage over X-rays in that they can be focused using electromagnetic lenses. This allows material characterizations at specific regions that are of real interest. Modern TEMs offer two methods to this end, including selected-area electron diffraction (SAED) and convergent-beam electron diffraction (CBED) techniques, respectively. In the SAED method, the diffraction pattern is confined to a selected area of the specimen via the employment of an aperture that is intuitively called the selected-area aperture (Fig. 3.12a). SAED can be performed on regions as small as 0.5 μm in diameter. By contrast, CBED technique provides the only access to the diffracting pattern from a region smaller than ca. 1 μm in diameter.

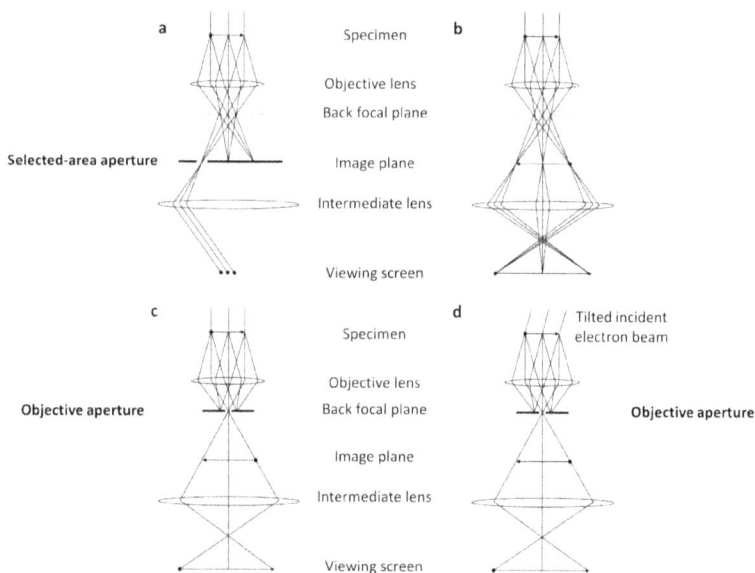

Figure 3. 12. (a) Selected-area electron diffraction mode, (b) apertureless imaging mode (c) bright field imaging mode and (d) axial dark field imaging mode in the TEM (*adapted from reference [110]*).

In addition to diffraction, the undiffracted and deflected electrons that successfully leave the thin specimen can alternatively be employed for real-space image formation with resolutions on the order of a few tenths to a few nanometres (Fig. 3.12b). Particularly, the TEM image formed upon exclusive collection of the undiffracted electrons is termed the bright field image (Fig. 3.12c). The background appears bright without the specimen in the bright-field TEM image. Contrarily, only the scattered electrons contribute to the dark field image formation and a black background is seen in the absence of the specimen (Fig. 3.12d). Experimentally, the transition between these two imaging modes is carried out via inserting an objective aperture at a specific location in the back focal plane of the objective lens. Noteworthily, additional caution must be exercised in the dark field imaging mode in view of the image-forming diffracted electrons traveling far from the optical axis (Fig. 3.12b). Such geometrical optics highly likely introduces large spherical aberration that is detrimental to the image resolution. This issue is successfully addressed in the TEM via tilting the incident electron beam by an angle equal to the characteristic diffraction angle of the specimen (Fig. 3.12d). In this way, the imaging electrons travel along the optical axis and pass through a centred aperture that ranges from ca. 5 to 20 μm in diameter.

There are three basic factors that account for the contrast level of a grayscale TEM photograph. For instance, when the primary electron beam passes through thick specimens or particular regions of a thin specimen, which are thicker or of higher atomic mass, strong incoherent elastic scattering takes place. In consequence, those regions show darker in the bright field TEM image. Such mechanism is termed mass-thickness contrast.[110,141] Mass-thickness contrast is based on incoherent elastic Rutherford scattering and takes place in amorphous materials and crystalline samples that are oriented in such a way that Bragg reflections are only weakly excited. Particularly, the quadratic dependence of the Rutherford scattering cross-section on the atomic number (Z) underlies the appearance of a Z-contrast image that is emphatically exploited in scanning transmission electron microscopy (STEM). Particularly, an annular detector is employed in the STEM mode of operation to collect these scattered transmitted electrons through relatively large angles. The Z-contrast imaging is highly useful in deriving compositional information, and moreover, a chemical map with a resolution of a few nanometres of the specimen with complex microstructures.

Significantly, when the specimen under investigation is a crystalline material, additional contrast is superimposed on the mass-thickness effect. This extra contrast builds on electron diffraction discussed in the foregoing paragraphs, wherein strong elastic scattering occurs when the crystallites in the specimen are well oriented to satisfy the Bragg condition. Given either undiffracted or diffracted electron beam underlying TEM imaging, the variation in diffraction intensity due to the presence of the lattice imperfections, e.g. dislocations, grain/phase boundaries and precipitates, is responsible for the contrast formation. Such diffraction contrast renders TEM a powerful tool not only for visualizing those crystallographic faults. Moreover, their visibility in the TEM image is employed to define the nature of those crystal defects.[102,110,141]

Noteworthily, when the zone axis of the grain in the crystalline specimen is parallel to the incident electron beam, many strong diffracted beams are produced (Fig. 3.11a). Moreover, if an apertureless imaging mode or a large objective aperture is employed (Fig. 3.12b), the undiffracted and several diffracted electron waves are allowed to mutually interfere and contribute to the image formation. In such a case, the resultant image contrast is dictated by the relative phases of the electron waves involved. This is in contrast to the mechanisms of aforementioned mass-thickness and diffraction contrasts, which employ exclusively the amplitude of the scattered electron beam, and is thereof termed phase-contrast imaging.[141] The interference pattern is manifested as a set of fringes in the image and ideally each dark fringe represents a lattice plane. In this way, the atomic structures of the materials are resolved and such structural image is called a high-resolution transmission electron microscope (HRTEM) image.

In this monograph, TEM was carried out to investigate the geometry and the morphology of topical Au@Nb@$H_xK_{1-x}NbO_3$ nanopeapods. An aberration-corrected FEI Titan[3] 80-300 microscope operated at 300 kV was employed to this end. The specimen was prepared via drop casting the Au@Nb@$H_xK_{1-x}NbO_3$ nanopeapods suspended in the hexane medium onto a 300 mesh copper grid with a carbon support film. The crystal structure and phase of the Au@Nb@$H_xK_{1-x}NbO_3$ nanopeapods were otherwise characterized using a Philips CM200 FEG/ST microscope operated at 200 kV. The chemical composition of the Au@Nb@$H_xK_{1-x}NbO_3$ nanopeapods was studied by high-angle annular dark-field scanning transmission electron microscopy (HAADF-STEM) combined with EDXS. The measurements were performed with a FEI Osiris ChemiSTEM microscope at 200 kV and equipped with a Bruker Quantax system (XFlash detector) for EDXS. The concentration profiles of different elements within the Au@Nb@$H_xK_{1-x}NbO_3$ nanopeapods were determined from the EDX spectra measured along a line-scan that passes through the center of a Au@Nb@$H_xK_{1-x}NbO_3$ nanopeapod. EDXS elemental maps of the Au@Nb@$H_xK_{1-x}NbO_3$ nanopeapods were additionally registered, which were subsequently employed to investigate the spatial distribution of diverse constituents. The maps were analyzed using the ESPRIT software (version 1.9) from Bruker.

4. Nanoengineerings for the Spikecube and Peapod Formation

Topical β-SnWO$_4$ spikecube and Au@Nb@H$_x$K$_{1-x}$NbO$_3$ nanopeapod artifacts throughout the dissertation were virtually prepared via a bottom-up approach, wherein the materials were built up either molecule-by-molecule from homogeneous building blocks or cluster-by-cluster from heterogeneous entities.[45,46] A polyol-mediated approach was employed for the β-SnWO$_4$ spikecube formation.[45] A soft-chemical multistep process was adopted to prepare the Au@Nb@H$_x$K$_{1-x}$NbO$_3$ nanopeapods (Fig. 2.11), which began with structural transformation of the precursor potassium hexaniobate (K$_4$Nb$_6$O$_{17}$) preformed in the solid state reaction into the tubular protonated mononiobate (H$_x$K$_{1-x}$NbO$_3$).[46] Heterogeneous integration and deposition of core-shell Au@Nb then followed up. The following section discusses the chemical background and the case studies, in which the nanoengineering was readily carried out via manipulating the reaction energetics and kinetics to derive those artifacts with the morphology that remarkably deviated from that favored by thermodynamics.

4.1 Polyol Synthesis

Polyol refers to multivalent alcohols, wherein several H-atoms of the hydrocarbon are substituted by hydroxyl functional groups (-OH).[143,144] This family begins with the ethylene glycol (EG) as the smallest prototype and comprises a large number of derivatives including diethylene glycol (DEG), triethylene glycol (TrEG) and polyethylene glycol (PEG), wherein those large molecules are mostly built up molecular-by-molecular from the EG monomer. In addition, propanediol (PDO), butanediol (BD), glycerol (GLY) and pentaerythritol (PE) likewise belong to this group. The presence of plural OH groups renders the boiling point of the polyols much higher than that of water, and moreover, allows most chemicals to be readily dissolved with a solubility comparable to that in water. Such nature renders these liquid-phase polyols highly promising as alternative mediators employed in the crystal growth at high temperature. Moreover, these functional groups endow the polyols with strong reducing power in terms of the facile oxidation of the alcohol to the aldehyde, the carboxylic acid, and eventually CO$_2$.[145] On such basis, the employment of polyols in the synthesis starts with the formation of elemental metals and alloys with high crystallinity. More importantly, the oxidative derivatives readily chelate the crystals via the carboxyl groups, which in turn quenches the access of the precursor ions to the solid nuclei effectively.[145] In this regard, the polyol synthesis is preferentially employed for the preparation of high quality small particles in the form of stable colloids with very minor agglomeration, wherein the particle size and shape are highly monodisperse and uniform, respectively. Such assets render this approach readily gained popularity in the fabrication of nanocrystalline metals, metal oxides, metal chalcogenides and non-metal elements.[143]

In this monograph, DEG with the characteristic boiling point up to 244 °C was employed in the synthesis of β-SnWO$_4$.[45,143] The reaction was carried out in a three-neck flask equipped with a reflux condenser and a magnetic stirring bar embedded in PYREX glass. 3.6 mmol of sodium tungstate (Na$_2$WO$_4$·2H$_2$O, 99%, Sigma-Aldrich) as the precursor were first dissolved in 150 mL of DEG at lightly elevated temperature of 60 °C. To this solution, 6 mL of an aqueous solution containing equimolar tin chloride (SnCl$_2$·2H$_2$O, 98%, Sigma-Aldrich) was added. Upon the injection, instantaneous complexation between Sn^{2+} and [WO$_4$]$^{2-}$ at the water/DEG interface occurred,

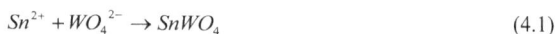

$$Sn^{2+} + WO_4^{2-} \rightarrow SnWO_4 \qquad (4.1)$$

leading to the precipitation of β-SnWO$_4$. This nucleated solid was subsequently passivated by the DEG molecules via surface adsorption. The overall system was then subjected to diverse thermal treatments with reaction temperatures of 120, 140, 160 and 180 °C, respectively. The overall reaction was carried out under nitrogen atmosphere for 1 h. Afterwards, centrifugation was employed to collect β-SnWO$_4$ that was further purified by repetitive dispersion/precipitation cycles with ethanol to remove DEG and excess precursors.

Given the synthesis performed in the context of a polyol-mediated approach, the crystal growth of β-SnWO$_4$ is mostly dictated by the majority DEG chelator. The coordinating behavior results in β-SnWO$_4$ prepared at 120 and 140 °C in the form of spherical nanoparticle with narrow size distribution that are validated by means of SEM and DLS, respectively (Fig. 4.1a,b). In contrast, the crystallinity of β-SnWO$_4$ is otherwise ruled by the thermal treatment. Particularly, the broadened Bragg peaks manifested in the diffraction pattern, which stems most likely from the significant overlap of characteristic diffraction lines of β-SnWO$_4$, suggests poor crystallinity (Fig. 4.1c,d). This issue is readily addressed via further raising the reaction temperature, as evidenced by the isolation of individual Bragg diffractions in the XRD patterns measured for β-SnWO$_4$ synthesized at 160 and 180 °C (Fig. 4.2a,b).

Surprisingly, the thermal treatment concurrently triggers morphologic transitions of β-SnWO$_4$, wherein spherical nanoparticles (insets in Fig. 4.1a,b) transforms into hexagonal microcubes enclosed by sharp crystal facets (Fig. 4.2c) that are further covered by a dense array of quasi-periodic nanotips (Fig. 4.2d). At the initial stage, the growth of the faceted β-SnWO$_4$

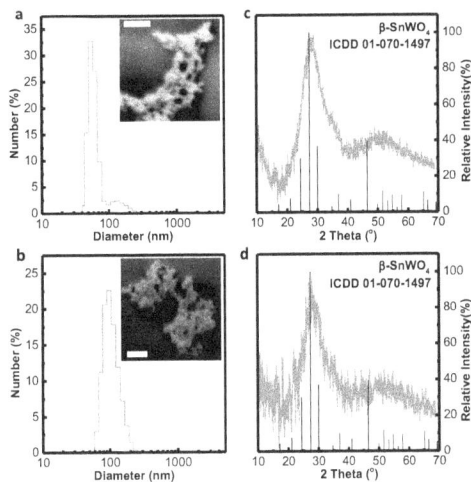

Figure 4. 1. (a,b) Size distribution histograms and (c,d) XRD patterns (reference: ICDD No. 1070-1497, β-SnWO$_4$) collected for β-SnWO$_4$ nanoparticles prepared at (a,c) 120 and (b,d) 140 °C, respectively. Insets: Representative SEM images are shown alongside (scale bar: 100 nm).

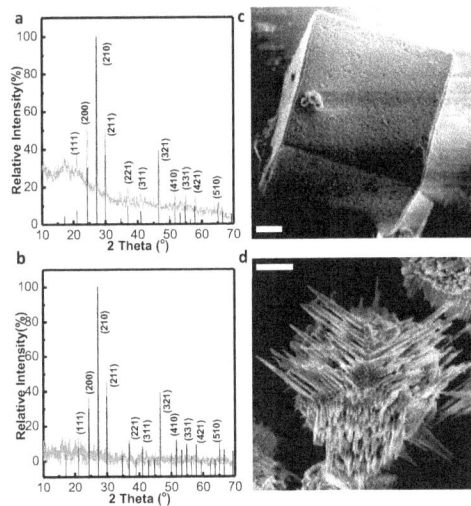

Figure 4. 2. (a,b) XRD patterns (reference: ICDD No. 1070-1497, β-SnWO$_4$) and (c,d) SEM images (scale bar: 2 μm) collected for β-SnWO$_4$ cubes and spikecubes prepared at (a,c) 160 and (b,d) 180 °C, respectively.

microcubes is carried out at the expense of the nanoscale colloids, which thus is regarded as an Ostwald ripening process. Particularly, the dissolution of highly energetic nanoparticles due to the appreciable surface-to-volume ratio is remarkably accelerated via the rise in temperature. The deposition rate of those redissolved particles on the solid nuclei that yet remains in the mother liquor depends otherwise on the facet exposed at the surface. In such context, the relative growth rate of different lattice planes primarily directs the final crystal shape, as illustrated in Fig. 4.3 that takes an imaginary two dimensional octagonal nuclei enclosed alternatively by two edges with one demonstrating higher growth rate than that of another as a simple example.[146] Evidently, the elongation of the solid edge with slow growth rate is carried out at the expense of the dot edge with faster growth rate, giving rise to a rectangular margined with the slow growing edges.

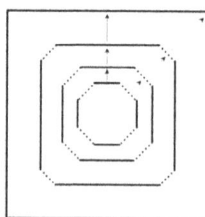

Figure 4. 3. Shape evolution during the nucleus growth of an imaginary two-dimensional seed. The length of the dot and slid arrow refers to the growth rate. Rapid addition to the dot edged leads to the elongation of the solid edges and the absence of the dot edges (*adapted from reference [146]*).

Given the starting β-SnWO$_4$ nanoparticles present in the three-dimensional space, the growth of these polyhedral seeds leads to the formation of hexagonal cubes. Specifically, this isometric geometry is in favor with β-SnWO$_4$, provided that the unit cell has been reported likewise in the shape of a cube.[147-149] In this connection, the side surfaces terminated by the low index {100} facets is concluded. Such exclusive coverage with a single set of planes over the crystal surface suggests an isotropic deposition rate of β-SnWO$_4$ to each facet of the solid hexahedra.[146] In consequence, these seeds grow in size over the reaction time whereas the shape changes no more, as manifested on SEM images (Fig. 4.4), wherein the majority of β-SnWO$_4$ crystallizes as cubes with the particle size falling in the range of ~2-9 μm.

In another run of synthesis at 180 °C, those cubes are further transformed into highly branched spikecubes (Fig. 4.2d) that are named for the first time in this study after the peculiar morphology, wherein numerous one-dimensional nanotips, nanopyramids and nanocones are built vertically on each side surface of the β-SnWO$_4$ cubes. Significantly, this anisotropic growth suggests the thermodynamically-driven Ostwald ripening process quenched at temperatures beyond 160 °C, which is validated by the cubic seed underlying the majority of spikecube bearing a characteristic size (d_c) consistent with that of the faceted cubes (Table 4.1). Moreover, the growth mechanism in such case is presumably under kinetic control, provided that the multiarmed shape is characterized by a larger surface area and consequently a higher surface energy that is disfavored by thermodynamics. Particularly, the dynamic water content plays a key role in such mechanistic transition, as elaborated below.

Figure 4. 4. SEM overview image of the β-SnWO$_4$ cubes prepared at 160 °C (scale bar: 10 µm).

Table 4. 1. Geometric features of the β-SnWO$_4$ spikecube.

Spike		
Arm length (h_l)	0.7 - 2 µm	
Base diameter (d_b)	200 nm	
Cube		
Particle size (d_c)	2 - 9 µm	

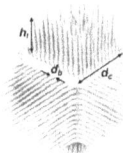

In a synthesis that is carried out under conditions similar to those employed in the preparation of β-SnWO$_4$, except for the exclusion of water, Na$_2$WO$_4$ is the major product (Fig. 4.5). This suggests that β-SnWO$_4$ that is the fundamental building block of a nucleus and a crystal is unavailable in the absence of H$_2$O. Such intimacy between H$_2$O and β-SnWO$_4$ renders the growth mechanism in the context of the polyol synthesis to be further hydrodynamically adaptable in view of the temporal and thermal fluctuations in the H$_2$O content during the reaction. In the infancy of the synthesis of the β-SnWO$_4$ spikecube, the presence of H$_2$O allows abundant β-SnWO$_4$ to be available to the birth of the nuclei that further grow into the crystals over the reaction time. During the temperature rising up to 160 °C, the crystal growth is likewise in the context of an Ostwald ripening process that can be accelerated by the thermal treatment, giving rise to a micrometer-sized cube. Such microcrystal otherwise denotes a substantial expense of the β-SnWO$_4$ in solution. Moreover, the reaction temperature is then raised to 180 °C that is nearly twice the normal boiling point (100 °C) of H$_2$O, presumably leading to only a trace level of H$_2$O yet present in the DEG pool. Overall, the rapid depletions of H$_2$O and β-SnWO$_4$ signify the growth rate to be significantly decelerated, and moreover, the growth mechanism transits from the thermodynamic to the kinetic mode when the diffusion rate turns out to be the limiting factor.[146,150,151]

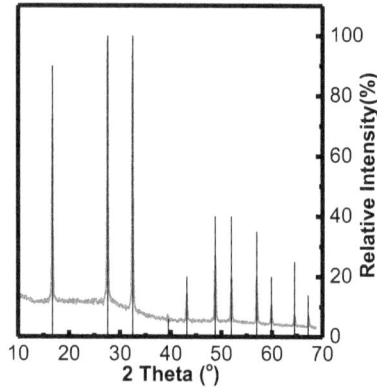

Figure 4. 5. XRD pattern of Na$_2$WO$_4$ collected in a polyol synthesis similar to that employed in the preparation of β-SnWO$_4$, except for H$_2$O being absent. (reference: ICDD-No. 0012-0772, Na$_2$WO$_4$).

The diffusion flux (J) of β-SnWO$_4$ impinging on the microcrystal facet is approximated by the Fick's law,

$$J = -D\nabla C_{\beta-SnWO_4} \qquad (4.2)$$

wherein $C_{\beta-SnWO_4}$ is the concentration of β-SnWO$_4$ in the majority DEG phase, $\nabla C_{\beta-SnWO_4}$ the concentration gradient over the diffusion layer at the microcrystal surface and D is the diffusion coefficient. Particularly, the precipitation of β-SnWO$_4$ from the DEG reservoir on the microcrystal approximately renders the concentration at the crystal surface to zero. Such premise leads to the magnitude of $\nabla C_{\beta-SnWO_4}$ mostly dictated by $C_{\beta-SnWO_4}$. Particularly, $C_{\beta-SnWO_4}$ is highly likely at a minimum level after the majority birth of β-SnWO$_4$ microcubes, which is validated by low aspect ratio between d_c and the geometry including the length (h_l) and the base diameter (d_b) of the anisotropic nanospikes building on the microcubic seed (Table 4.1). On such basis, minuscule $\nabla C_{\beta-SnWO_4}$ is concluded. Moreover, the well-known Stokes-Einstein relation (Eqn. 3.17) suggests

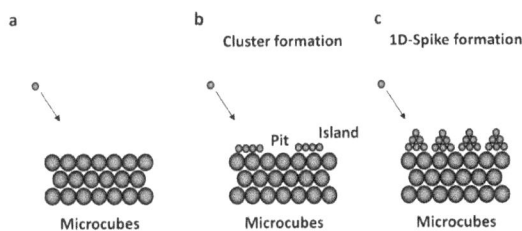

Figure 4. 6. Schematic illustration of the growth mechanism responsible for the transformation from a faceted β-SnWO$_4$ microcube to a β-SnWO$_4$ spikecube. (a) Diffusion of β-SnWO$_4$ from the DEG pool to the cubic microcrystals at low flux. (b) Deficient β-SnWO$_4$ feed leaded to the deposition in the form of discrete clusters in preference to homogeneous monolayer. (c) Shadowing effect renders the incoming β-SnWO$_4$ hardly available to the concave pit, giving rise to the voids distributed in the nanospike matrix.

Figure 4. 7. SEM top view image of a β-SnWO$_4$ spikecube (scale bar: 5 μm).

that D depends highly on the temperature and viscosity of the solution, and the hydrodynamic size of the β-SnWO$_4$ nucleus. Particularly, the majority DEG is advantageous due to the high viscosity that is more than 30 times that of H$_2$O, very likely leading to small D that further reduces J.[143] Taken together, the diffusion of β-SnWO$_4$ from the DEG pool to the cubic microcrystals, which is one of the kinetic elementary steps in the overall deposition scheme, turns out to be the rate determining step.

Particularly, exceedingly low J results in precipitated β-SnWO$_4$ on the facets of the underlying hexahedral microcube hardly meeting each other, which in turn leads to the deposition in the form of a few tiny clusters in lieu of a homogeneous monolayer. In consequence, the crystal surface becomes progressively rough and the shadowing effect of such undulation leads to protrusive islands in place of concave pits readily capturing the impinging Sn^{2+} and WO_4^{2-} ions (Fig. 4.6). Such scheme is well reinforced by the presence of numerous voids within the nanospike matrix, which is reminiscent of the infant cavities (Fig. 4.7). Moreover, this mechanism is otherwise consolidated by the excellent agreement in the one-dimensional geometry of the nanospike with that reported in earlier studies on the chemical deposition in the kinetic columnar growth mode, wherein the limited diffusion rate of the adatoms otherwise results in the cluster formation on the substrate.[150]

In summary, fine β-$SnWO_4$ particles with the dimension over nano- and microscopic scales and additional facets and morphological engineerings are readily carried out in the context of the polyol synthesis (Fig. 4.8). Moreover, the currently developed growth scheme for the first time reveals a powerful effect of the thermal treatment on the crystal quality, including not only the well-documented crystallinity enhancement but also a crystal form manipulation. Last but not least, the present work is believed to successfully provide a brand-new insight into the polyol-mediated crystallization, further reinforcing the knowledge of material preparation.

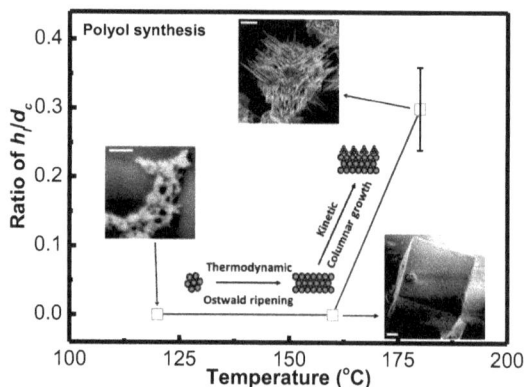

Figure 4. 8. Size and morphological evolution of β-$SnWO_4$ as a function of reaction temperature in the context of the polyol synthesis. Representative SEM images of specific particles in distinct runs of synthesis are shown alongside.

4.2 Soft-chemical Solid-state Retrosynthesis

Solid-state inorganic chemistry thrives on the rich library of solids that can be formed using a wide variety of synthetic approaches, including the simplest solid-solid reaction at high temperature and the versatile hydrothermal and sol-gel reactions at moderate temperature.[152-156] Particularly, the solid-state approach provides the most straightforward route to diverse new solids, which is thus prevalently employed in the material preparation. Nevertheless, such ordinary scheme implies the least control over the synthesis, giving rise to most solids that are thermodynamically favorable.[41,72] This connotation thus calls for a secondary engineering on those preformed solids in order to yield additional metastable alternatives that are preferentially favored by kinetics.[72] This gives birth to the soft-chemical solid-state synthesis that builds on the low-temperature modification of solid precursors.[72,153,157-165] This moderate thermal treatment effectively quenches extensive bond breakings and rearrangement of the structural framework, leading to the final product structurally approximating the starting materials. In this connection, the soft-chemical solid-state approach is retrosynthetic in nature and most treatments involve metathesis reactions to interchange weakly bonded cations and anions in the framework structure. On such basis, the solid-state inorganic retrosynthesis is extensively performed on the layered materials with a particular interest in the perovskites (Fig. 2.10).[72,74,157] In consequence, a host of new compounds with the crystal structure and morphology reminiscent of the lamellar framework of the precursors are successfully designed.

In this dissertation, potassium hexaniobate ($K_4Nb_6O_{17}$) characterized by a layered crystal structure was employed as the starting material in the soft-chemical solid-state retrosynthesis of $H_xK_{1-x}NbO_3$.[41,46,157,166] The overall reaction began with the preparation of this layered precursor, wherein reagent-grade potassium carbonate (K_2CO_3, 99.99%, Carl Roth) and niobium pentoxide (Nb_2O_5, 99.99%, Sigma-Aldrich) with a molar ratio of 1.1:1.5 were first ground together and heated in an alumina crucible at 900 °C for 1 h before continuing to 1050 °C for another 24 h.[46,157] The cooling product was washed three times with distilled water and acetone respectively and dried in the oven overnight. The purified solid was then immersed in warm (60 °C) and concentrated (3 mol L^{-1}) hydrochloric acid (HCl) for at least four days.[46,157] After the acid-treatment, the product was centrifuged and washed with distilled water and acetone, three times in each case, and dried overnight. Eventually, this acid-treated solid (0.5 g) was mixed with 0.75 g of tetrabutylammonium (TBA) hydroxide 30-hydrate (TBAOH·30H$_2$O, $C_{16}H_{37}NO·30H_2O$, 98%, Sigma-Aldrich) and 25 ml of oleylamine (OAm, $C_{18}H_{37}N$, 70%, Sigma-Aldrich) in 40 mL of toluene (C_7H_8, 99.5%, Thermo Fisher Scientific).[46,167] This mixture was magnetically stirred at ambient temperature for 1 h before transferring into a Teflon-lined stainless steel autoclave (Parr, model 4590, 100 ml). The overall system was then heated to 220 °C with a ramping rate of 2 °C min^{-1} and the reaction was carried out at 220 °C for 6 h. The final product was collected by centrifugation, washed with ethanol several times and then dried overnight.

The molar ratio of K_2CO_3 and Nb_2O_5 employed in the initial solid state reaction is straightforwardly dictated by the composition stoichiometry of $K_4Nb_6O_{17}$. A slight excess of K_2CO_3(10 mol %) aims at remedying for the material volatilization during annealing to ensure the phase purity of the solid precursor.[157] The intense (020) and (040) Bragg peaks at 2θ of ca. 5° and 10° in the collected XRD pattern (Fig. 4.9a) validates the layered crystal structure of $K_4Nb_6O_{17}$, wherein the basal lattice planes, (-Nb_6O_{17}-)$_n$, stacks along the b-axis (Fig. 4.9b). The formation of $K_4Nb_6O_{17}$ hydrate ($K_4Nb_6O_{17}\cdot3H_2O$) stems most likely from the reversible absorption of aerial moisture upon the exposure to the atmosphere.[168]

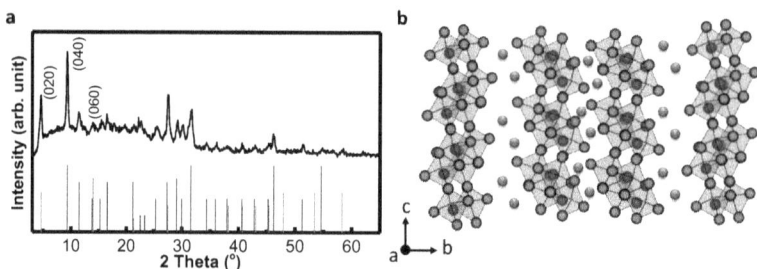

Figure 4. 9. (a) XRD pattern collected for $K_4Nb_6O_{17}$ synthesized in the context of a solid-state reaction. (reference: ICDD-No. 0021-1297, $K_4Nb_6O_{17}\cdot3H_2O$ (blue line); ICDD-No. 0014-0287, $K_4Nb_6O_{17}$ (green line)). (b) Crystal structure of $K_4Nb_6O_{17}$.

The soft-chemical treatment of $K_4Nb_6O_{17}$ starts with the ion-exchange in acid, wherein the interlayer potassium ions (K^+) serving as the zippers bonding massive (-Nb_6O_{17}-)$_n$ sheets together are substituted for protons (H^+).[72,157] This soft chemistry turns $K_4Nb_6O_{17}$ into the protonated form ($H_xK_{4-x}Nb_6O_{17}$) that is a solid acid, which is in favor of subsequent intercalation of basic TBAOH.[72] Noteworthily, the subsequent intercalation in this study is carried out in nonpolar toluene that highly likely suppresses the dissociations of the crystal water and hydroxide (OH^-) from TBAOH.[46,167] This in turn results in this intercalates sterically bulky, which most likely pack loosely in the interlamellar sites, leading to the interlayer gallery readily swelling up with the solvent molecules.[72] As soon as the van der Waals forces between each basal lattice are overcome, $H_xK_{4-x}Nb_6O_{17}$ delaminates and a colloidal suspension of exfoliated (-Nb_6O_{17}-)$_n$ sheets is formed. Significantly, (-Nb_6O_{17}-)$_n$ sheets are characterized by the lack of inversion symmetry. In other words, one side of the (-Nb_6O_{17}-)$_n$ sheet is sterically more shielded than another side (Fig. 4.10a,b), which turns out to render the lamella inherent mechanical strain.[157] The isolated (-Nb_6O_{17}-)$_n$ sheets release such strain upon irreversibly coiling away from the shielded side, giving rise to tubular in lieu of lamellar conformation (Fig. 4.10c). On such basis, this spontaneous nanoorigami is highly likely entropy-driven and the negatively charged anionic surface of the $(-Nb_6O_{17}-)_n$ sheet is electrically neutralized by the adsorption of K^+ and H^+.[157]

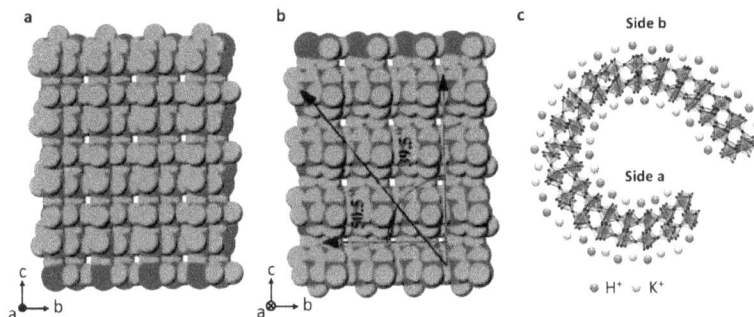

Figure 4. 10. Space-filling model of single $(-Nb_6O_{17}-)^{4-}$ sheet viewed along the *a* directions (a) out of and (b) into the page, respectively (*adapted from reference [157]*). The cations are omitted for clarity. (c) The atomic density at side (b) is higher than at side (a), leading to the scrolling of the $(-Nb_6O_{17}-)^{4-}$ sheets.

Particularly, the retrosynthesis from $H_xK_{4-x}Nb_6O_{17}$ to $H_xK_{1-x}NbO_3$ takes place presumably in parallel with the morphological transformation via a hydration process with the coordinated crystal water of $TBAOH \cdot 30H_2O$ during the intercalation, as validated by quantitative EDXS analysis (Fig. 4.11a).[46,166]

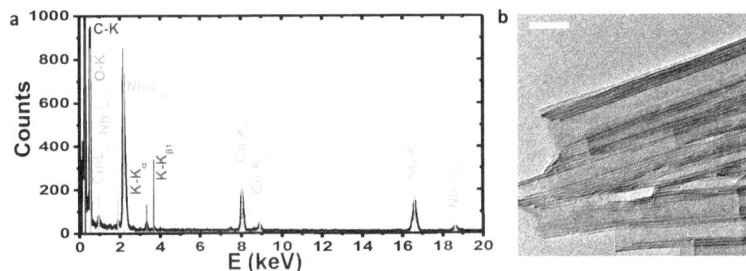

$$H_x K_{4-x}Nb_6O_{17} + H_2O \rightarrow 6H_xK_{1-x}NbO_3 \qquad (4.3)$$

Figure 4. 11. (a) EDX spectrum and (b) TEM image (scale bar: 70 nm) ascribe a stoichiometry of $K_{0.03}NbO_3$ and a tubular form to the final product. Given H undetectable by EDXS, a real composition of $H_{0.97}K_{0.03}NbO_3$ is concluded.

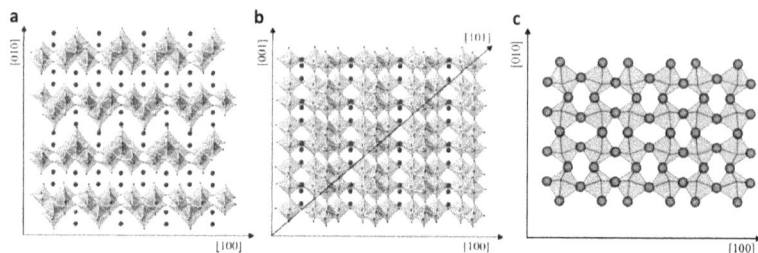

Figure 4. 12. Crystal structure of layered $K_4Nb_6O_{17}$ along the (a) [001] and (b) [010] direction, respectively (*adapted from reference [166]*). (c) Crystal structure of cubic $HNbO_3$ perovskite. The cations are omitted for clarity.

$H_xK_{1-x}NbO_3$ yet adopts the anisotropic tubule shape of the $H_xK_{4-x}Nb_6O_{17}$ precursor, as manifested by TEM images (Fig. 4.11b). Such crystal form is thermodynamically disfavored by $H_xK_{1-x}NbO_3$ in view of the characteristic isometric cubic perovskite structure (Fig. 4.12c). This conclusion in turn reiterates that the soft-chemical solid-state retrosynthesis is carried out under kinetic control.[72] Particularly, $(-Nb_6O_{17}-)_n$ sheets consist of a central layer of $[NbO_6]$ octahedra shearing corners in the [100] direction and alternating edges and corners in the [001] direction (Fig. 4.12a,b).[166,169] Such corner-shearing $[NbO_6]$ octahedra are also an inherent structural feature of $H_xK_{1-x}NbO_3$ (Fig. 4.12c).[170] This crystal analogy allows $H_xK_{1-x}NbO_3$ substantially inheriting the framework of $H_xK_{4-x}Nb_6O_{17}$, which in turn directs the final crystal form.

In summary, tubular $H_xK_{1-x}NbO_3$ nanoscrolls are successfully synthesized by the soft-chemical solid-state retrosynthesis from $H_xK_{4-x}Nb_6O_{17}$. Evidently, the anisotropic shape is thermodynamically disfavored by $H_xK_{1-x}NbO_3$ bearing an isometric cubic structure, which in turn substantiates the retrosynthesis preferentially occurring under kinetic control. Particularly, the grain orientation of the retrosynthetic $H_xK_{4-x}Nb_6O_{17}$ precursor mostly dictates the final crystal form of $H_xK_{1-x}NbO_3$.

4.3 Underpotential Deposition (UPD)

Underpotential deposition (UPD) refers to the deposition of monolayered metal adatoms onto a foreign metal substrate at reduction potentials being positive in relation to the characteristic reversible Nernst potential.[171-173] Such potential shift is virtually of relevance to the minor energy barrier of the heterogeneous deposition with respect to homogeneous nucleation.[171] Given the underpotential shift in general quantitatively characterized by means of transient electrochemical techniques such as cyclic voltammetry,[171-173] early investigations on UPD particularly aim at correlating such phenomenon with the physiochemical properties of metallic deposit and electrode. Particularly, a physical model building on the work function difference between the metals well interprets the underpotential shift.[171] Specifically, partial charge transfer stemming from the work function difference renders the chemical bond between the substrate and adsorbate the ionic character. Particularly, the electron is exclusively transferred from the metal deposit to the electrode with the extent mostly dictated by the work function difference.[171] The chemical bond thus gains a polarity in the electron distribution, which accounts primarily for the favorable deposition of metal adsorbates onto an alien substrate.

This model was very recently extrapolated to the bimetallic nanocrystal growth and employed predominantly in the synthesis of core-shell heterostructures that strongly demand a monolayer accuracy of the shell thickness.[174,175] The overall process virtually involves UPD coupled with seed-mediated growth and/or galvanic replacement reactions, wherein a seed-mediated approach aims at carrying out the bimetallic nanocrystal growth in a stepwise mode.[174] The process begins with preforming the nanocrystals of the core metal, which in turn serve as the primary sites for the subsequent deposition of the shell metal. In this way, heterogeneous deposition is exclusively carried out in the absence of homogenous nucleation of the shell metal, leading to a remarkable uniformity in size, shape, composition and structure of the final bimetallic nanocrystals. Moreover, the deposition of the shell metal performed at underpotentials leads to a very conformal coating with the thickness of only single or very few atomic layers.[174,175] Particularly, the galvanic replacement process allows this UPD skin exchangeable by alternative metals by means of the electrochemical redox reaction, which further reinforces the compositional and structural sophistications of the bimetallics.[174] To meet this end, the substituent must have a stronger oxidizing power than that of the oblation. More intriguingly, repeating the UPD and galvanic replacement sequentially allows the formation of multilayer core-shell heterostructures mimicking the pattern of an onion or matryoshka.[174]

In this monograph, bimetallic core-shell Au@Nb nanocrystals were synthesized via preforming spherical Au nanocrystals that serve as the seeds for the deposition of Nb at underpotentials.[46,176] In a standard synthesis, the retrosynthetically preformed $H_xK_{1-x}NbO_3$ nanoscrolls (20 mg), 10 mg of gold chloride tetrahydrate ($HAuCl_4 \cdot 4H_2O$, 99.9%, Sigma-Aldrich), 160 μL of oleic acid (OAc, $C_{18}H_{34}O_2$, 90%, Sigma-Aldrich), and 165 μL of OAm were added to 3 ml of hexane (C_6H_{14}, 99%, Thermo Fisher Scientific). The reaction system was then vigorously stirred and heated to nearly 60 °C for 24 h before cooling to room temperature. The final product was purified by repetitive dispersion/precipitation cycles with ethanol and finally dispersed in either deionized water or hexane for storage. Au nanocrystals as the seeding nuclei employed in the UPD-coupled seed-mediated approach is prepared using common solution reduction techniques, wherein OAm and OAc are employed as the reducing agents.[46,176] Noteworthily, in this study the spatial distribution of OAm over the reaction system is highly inhomogeneous. Given the preliminary addition of OAm in the soft-chemical retrosynthesis, OAm very likely saturates the cavity of the $H_xK_{1-x}NbO_3$ nanoscrolls during the nanoorigami process described in the preceding unit (Fig. 4.10). This in turn kinetically accelerates the nucleation rate of Au nanocrystals in view of the highest collision frequency between OAm and the $AuCl_4^-$ precursor therein. In consequence, most Au nanocrystals seed inside the $H_xK_{1-x}NbO_3$ nanoscrolls, as manifested in the TEM image (Fig. 4.13).

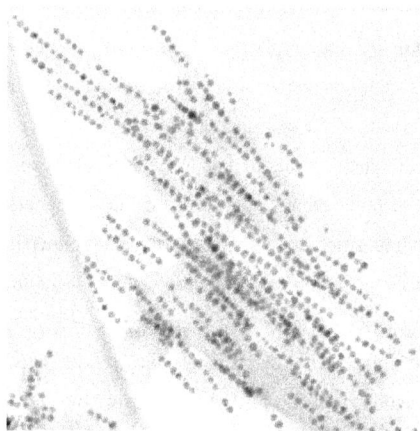

Figure 4. 13. TEM image validated that most Au nanocrystals unidirectionally seeded inside the cavity of the retrosynthetically preformed $H_xK_{1-x}NbO_3$ nanoscrolls (scale bar: 100 nm).

Particularly, discrete Au nanocrystals with highly uniform nanometric breaks of ca. 2 nm suggest that the nanocrystal surface is functionalized by the oxidized OAm molecules. This is in excellent agreement with the reports in the literature that OAm usually plays a dual role in the nanocrystal growth as the surfactant in addition to the reducing agent.[174] Such conclusion is further reinforced by the interparticle distance that matches satisfactorily with the chain length of the OAm molecule.[177] Statistical analysis over hundreds of TEM images reveals that these Au nanoseeds crystallize into a penta-twinned decahedral structure, wherein five single-crystal, tetrahedral domains join together at five twin boundaries by sharing one of the edges along a 5-fold axis (Fig.

4.14). This is the thermodynamically stable internal structure for Au nanocrystal with intermediate size between 3 to 15 nm, agreeing well with another statistical conclusion from the TEM images that most Au nanocrystals in this study are sub-10 nm sized.[178]

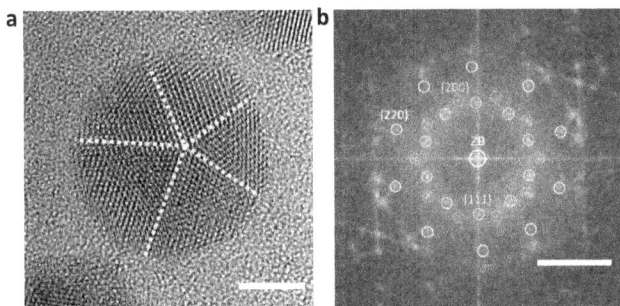

Figure 4. 14. (a) HRTEM image (scale bar: 3 nm) and (b) diffractogram (scale bar: 6 nm^{-1}) of the Au nanocrystal in (a) with calculated diffraction pattern with Miller indices for a Au decahedron along its fivefold symmetry axis (Au with space group *Fm-3m*, a = 4.08 Å). Orange dashed lines in (a) delineate five twin boundaries between individual single-crystal domains (lines are guide to the eye). The white circle in (b) indicates the zero-order beam (ZB).

Noteworthily, subsequent UPD-coupled seed-mediated approach for the bimetallic core-shell Au@Nb nanocrystal preparation is carried out in the same chemical batch in this study, wherein the Au nanocrystals are produced. More importantly, those preformed Au nanoseeds are in close proximity to the $H_xK_{1-x}NbO_3$ nanoscrolls, which is employed exclusively as the Nb^{5+}-containing precursor, in a biomimetic peapod configuration (Fig. 4.13). The biomimicry of this peapod scheme likewise promotes the collision frequency between the Au nanocrystals and Nb^{5+} that leaks out of the $H_xK_{1-x}NbO_3$ nanoscrolls presumably during the phase transition, provided that $H_xK_{1-x}NbO_3$ contains a lower stoichiometric content of Nb than that of $H_xK_{4-x}Nb_6O_{17}$. In this way, UPD is kinetically favored and feasible, provided that Nb^{5+} cannot be directly reduced by either OAm, a well-known mild reducing agent, or Au. UPD in this study succeeds in Nb^{5+} first absorbed on the surface of the Au nanocrystal, sharing a small portion of the free electron cloud through the empty orbital.[179] This leads to the electron cloud being spherically distributed around Au turning into an elliptical distribution (Fig. 4.15).

This perturbation in electron distribution gives rise to a partial positive charge on the Au surface, which is subsequently neutralized by the electron transfers from OAm. In other words, Au serves as the electron relay in this carrier delivery, which in turn satisfies the premise that Nb^{5+} is

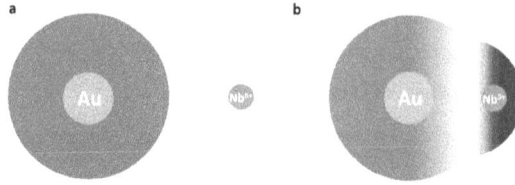

Figure 4. 15. Free electron cloud distribution around Au nuclei (a) before and (b) after Nb^{5+} absorption (*adapted from reference [179]*).

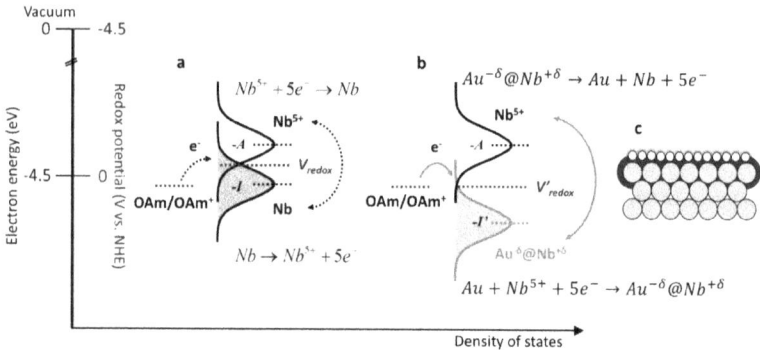

Figure 4. 16. (a) Thermodynamics disfavors the homogeneous reduction of Nb^{5+} by OAm, but (b) the presence of Au nuclei presumably tuned such process thermodynamically feasible via a UPD route due to (c) the partial charge transfer from Nb adatoms (cyan spheres) to Au nanocrystals (orange spheres). Moreover, the particle positive charge (green cloud) present at Nb adatoms likely stimulated the deposition of next few atomic layers likewise at underpotentials.

reduced at underpotentials by the electron transporting through the nanocrystalline Au substrate.[171-173] The elementary steps making up the UPD of Nb

$$Au + Nb^{5+} + 5e^- \leftrightarrow Au@Nb \tag{4.4}$$

involve,

$$Au + Nb^{5+} + 5e^- \rightarrow Au^{-\delta}@Nb^{+\delta} \tag{4.5}$$

$$Au^{-\delta}@Nb^{+\delta} \rightarrow Au + Nb^{5+} + 5e^- \tag{4.6}$$

wherein the reduction and oxidation potentials are of relevance to the summary of first to fifth ionization potential (I) of Nb and first to fifth electron affinity (A) of Nb^{5+}, respectively.

Particularly, the large gap in terms of work function between Nb (4.3 eV) and Au (5.1 eV) results in partial electron transfer (δ) from Nb to Au, which in turn increases the ionization potential of the Au$^{-\delta}$@Nb$^{+\delta}$ deposit (Eqn. 4.6 and Fig. 4.16). In consequence, the redox potential, which in terms of a Fermi distribution function corresponds to the energy state with a 50% probability of electron occupancy[47,48], positively shifts (on scale of V vs. NHE) with the quantity dictated by

$$\Delta U_p = \alpha \Delta \Phi \qquad (4.7)$$

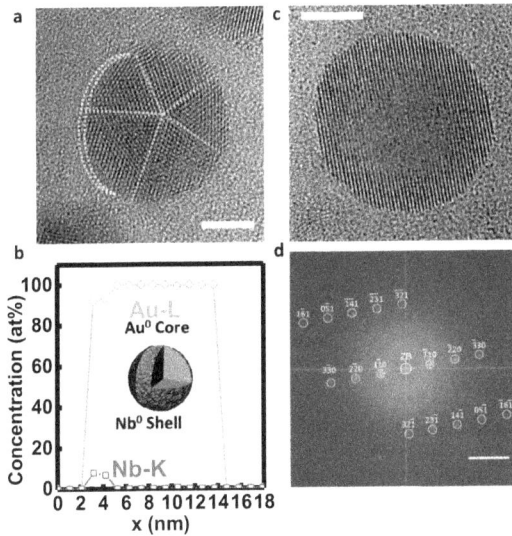

Figure 4. 17. (a) HRTEM image (scale bar: 3 nm) and (b) corresponding EDXS line profile substantiated a monolayer deposition of Nb shell onto nanocrystalline Au decahedron at underpotentials. Orange and sky-blue dashed lines in (a) delineate the penta-twinned internal structure of underlying Au and the boundary between Au and Nb, respectively (lines are guide to the eye). (c) HRTEM image (scale bar: 4 nm) validated further UPD of the next few atomic Nb layers. (d) Diffractogram (scale bar: 7 nm^{-1}) of the conformal core-shell Au@Nb nanocrystal in (c) and calculated diffraction pattern with Miller indices of bulk cubic Nb (space group *Im-3m*, a = 3.32 Å) in the [115]-zone axis. The white circle in (d) indicates the zero-order beam (ZB).

In Eqn. 4.7, ΔU_p is the underpotential shift, α the proportional coefficient of 0.5 V eV^{-1} and $\Delta \Phi$ represents the work function difference between two metals.[171] In this study, ΔU_p amounts to 0.4 V that highly likely turns the reduction of Nb^{5+} by the Au-absorbed-OAm that is otherwise thermodynamically feasible.

In consequence, a monolayer Nb deposit builds upon the top of the Au nanocrystals, as evidenced by the EDXS line scans (Fig. 4.17a,b). More importantly, partial positive charge on the

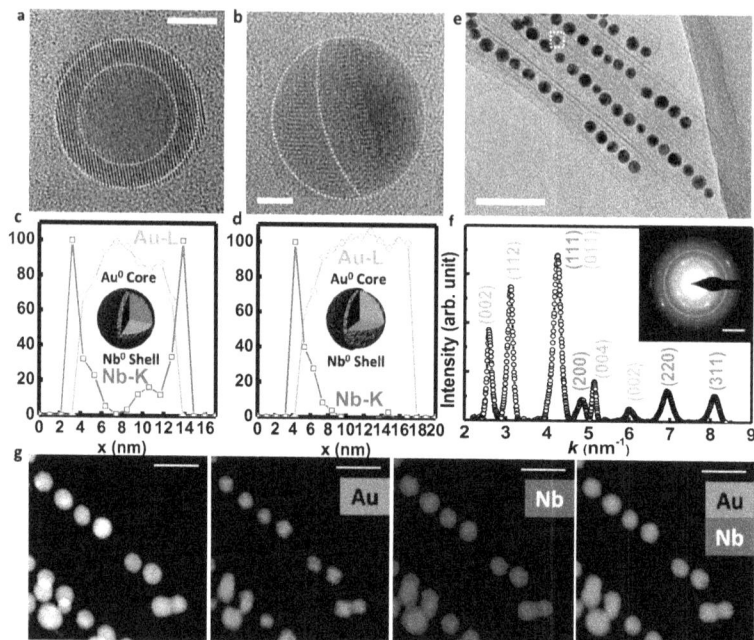

Figure 4. 18. (a,b) HRTEM images and (c,d) corresponding EDXS line profiles (element contributions from the $H_xK_{1-x}NbO_3$ nanoscrolls were subtracted) of the bimetallic core-shell Au@Nb nanocrystals bred either (a,c) in or (b,d) atop (yellow framed region in (e)) the $H_xK_{1-x}NbO_3$ nanoscrolls (e) in a peapod-like configuration. The formation of either (a,c) conformal (scale bar: 4 nm) or (b,d) partial Nb bridge (scale bar: 2 nm) depends on the spatial distribution of Au nanocrystals with respect to the $H_xK_{1-x}NbO_3$ nanoscrolls. (e,f) Intense (002), (112) and (004) diffraction peaks manifested in the (f) azimuthally averaged SEAD pattern that was derived from the Debye-Scherrer rings in the initial SAED pattern (inset in (f), scale bar: 3 nm^{-1}) suggested a (00l)-preferred orientation of HNbO$_3$ with a cubic structure, which was ascribed to the tubular shape of $H_xK_{1-x}NbO_3$ nanoscrolls evidenced in the (e) TEM image (scale bar: 50 nm). (g) High-angle annular dark-field scanning TEM (HAADF-STEM) image and EDXS elemental maps (scale bars: 20 nm) punctuated the formation of bimetallic core-shell Au@Nb nanocrystals (lilac) upon the concomitant distribution of Au (red) and Nb (blue) elements.

Nb adatoms (Fig. 4.16c) presumably stimulates the next few atomic layers of Nb to likewise deposit at underpotentials, leading to a monocrystalline, very thin and conformal Nb shell atop the nanocrystalline Au decahedron, as manifested in the TEM image and electron diffractogram (Fig. 4.17c,d).

Statistical analysis reveals that the form of the bimetallic core-shell Au@Nb nanocrystals depends strongly on the spatial distribution of the Au nanocrystals with respect to the $H_xK_{1-x}NbO_3$ nanoscrolls in the UPD-coupled seed-mediated growth (Fig. 4.18e). Particularly, a conformal Nb shell with thickness of ca. 1 nm that corresponding to atomic layer number less than six (1.5 nm thick)[174] is atop most of Au nanocrystals preformed inside the $H_xK_{1-x}NbO_3$ nanoscrolls (Fig. 4.18a,c,f,g). In contrast, a partial Nb bridge builds preferentially on a few Au nanocrystals scattering over the $H_xK_{1-x}NbO_3$ nanoscrolls (Fig. 4.18b,d,e).

In summary, an one-pot approach involving two-phase synthesis, which begins with a self-nucleation process of the noble Au metal that in turn serves as the seeds for heterogeneous deposition of the transition Nb metal, is developed for the formation of bimetallic core-shell Au@Nb nanocrystals. Moreover, monodisperse Au@Nb nanocrystals with sub-10 nm size are engineered into a biomimetic peapod pattern, wherein discrete core-shell bimetallics unidirectionally seed inside the retrosynthetically preformed $H_xK_{1-x}NbO_3$ nanoscrolls with nanoscale resolution over sub-microscopic distance. Last but not least, the present contribution is believed to successfully provide a brand-new insight into the UPD-coupled seed-mediated crystallization, further reinforcing the knowledge of bimetallic nanocrystal growth.

5. Nanoarchitecture-mediated Photoelectrocatalytic Functionality

Insofar, progress is made in illuminating the synthetic mechanisms of engineering β-SnWO₄ and Au@Nb@H$_x$K$_{1-x}$NbO₃ into the spikecube and peapod configurations. In the following section, efforts are geared toward studying the significance of such material nanoarchitectures to the photocatalytic and photoelectrocatalytic functionalities. Particularly, photocatalytic dye degradation for water remediation and photoelectrochemical water splitting for fuel generation are employed in this dissertation as the evaluation metrics.[41,180] Given that the prefix "photo-" literally suggests that heterogeneous photoelectrocatalytsis starts exclusively with light harvesting by the materials, the photoabsorption ability of the topical artifacts is first studied. The suffix "-electrocatalysis" otherwise expresses that redox reactions are afterwards triggered by the photogenerated charge carriers on two topical semiconducting catalysts. Given the reducing and oxidizing powers of the electrons and holes determined primarily by the energetics of CB and VB, the characteristic band structures intimately dictated by the atomic coordination of two topical chemical systems are further explored. Moreover, action spectra are utilized to trace the optical dependency of the performance in order to exclude undesirable paradoxes from the estimation to ensure the scientific validity.[180] Likewise, the reaction products are additionally characterized to reinforce the reliability.[41]

5.1 Photocatalytic Dye Degradation Metrics

In this metrics, the consumption course of two model dyes including cationic methylene blue (MB) and zwitterionic rhodamine B (RhB) during irradiating spikecubic β-SnWO₄ photocatalyst is registered.[181,181] The measurement was carried out in a DURAN glass reaction cell containing a suspension of powdery photocatalyst (0.3 g L⁻¹, 12 mg) under investigation in an aqueous dye solution (5.44 mg L⁻¹, 40 mL) with continuous agitation using a magnetic stirrer. Before illumination, the suspensions were magnetically stirred in the dark for 1 h to ensure the establishment of an adsorption/desorption equilibrium of the dyes on the photocatalyst surface. During photoirradiation, 2 ml of aliquot were sampled from the reaction reservoir every 10 minutes. Centrifugation was applied to separate the photocatalyst from the dye solution that was then analyzed via UV-VIS spectrophotometry. The concentration evolution of the dye solution was derived from the absorbance variation registered at the signature wavelengths of 665 and 554 nm for MB and RhB, respectively (Fig. 5.1a). Significantly, monochromatic light with featured wavelength of 366 nm was utilized as the illumination source. Such scheme circumvents the competition between model dye and the β-SnWO₄ spikecube for incident photons in terms of the minor overlap of respective absorption bands, as manifested in the UV-VIS absorption spectra (Fig. 5.1a). Moreover, the incoming photons have an energy exceeding the band gap of the β-SnWO₄ spikecubes, which is 3.01(±0.03) eV (Fig. 5.1b) derived from the Kubelka-Munk transformed diffuse reflection spectrum (Fig. 5.1a) using the Tauc law (Eqn. 3.13). Such scheme allows to reliably ascribe the decomposition of dye molecules to the photocatalytic effect of the β-SnWO₄ spikecubes. This is further reinforced by the negligible self-photolysis of the dye molecules under similar conditions except for the absence of the β-SnWO₄ spikecubes (Fig. 5.2).

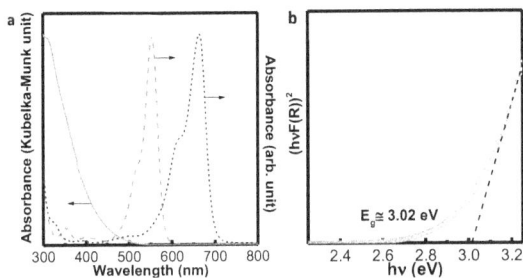

Figure 5. 1. (a) UV-VIS absorption spectra (right axis) of aqueous RhB (brown dash line) and MB (blue dot line) solutions along with the Kubelka-Munk transformed diffuse reflectance spectra (left axis) of the β-SnWO₄ spikecubes (red solid line). (b) Tauc plot (for direct interband transition) of the β-SnWO₄ spikecubes.

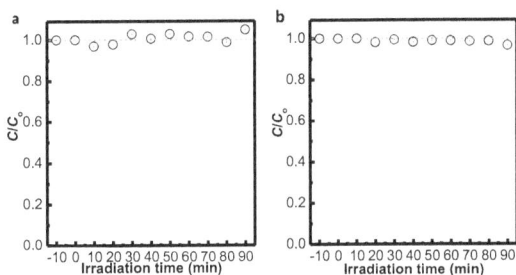

Figure 5. 2. Negligible attenuation in concentration (C) normalized to that (C_0) before illumination of aqueous (a) MB and (b) RhB solutions under monochromatic light illumination (wavelength: 366 nm) at the absence of spikecubic β-SnWO₄ photocatalyst.

On such basis, normalization of the molar amount of the decomposed dye to that of the β-SnWO₄ spikecubes, which gives rise to the turnover number (TON),

$$TON = \frac{Mole\ of\ reacted\ molecules}{Mole\ of\ atmos\ in\ the\ photocatalyst} \qquad (5.1)$$

is employed in registering the reaction course (Fig. 5.3a-c).[41] Significantly, TON measured for the β-SnWO$_4$ spikecubes in the studied period exhibits a linear dependency on the photoirradiation time, suggesting an independent degradation rate (r_{ph}) of the dye concentration (C_{dye}).

In general, r_{ph} of a heterogeneous photochemical reaction that virtually builds on a monomolecular surface reaction mechanism is formulated as[180]

$$r_{ph} = k' \theta_s C_s \tag{5.2}$$

In Eqn. 5.2, k' is the rate constant, C_s the adsorption capacity of solid photocatalyst and θ_s is the surface coverage with the reactant. Particularly, k' can be further deconvoluted into,

$$k' = \frac{I\Phi_{ph}k_{red}}{k_r} \tag{5.3}$$

wherein I is the incident light flux that can be regarded as a constant throughout this study. In contrast, Φ_{ph} is of relevance to the absorption efficiency of the photocatalyst to the light, k_{red} and k_r represent interfacial reaction rate constant of photogenerated electron-hole pairs with surface-adsorbed substance and competitive recombination rate constant, respectively. Evidently, these terms depend strongly on the physicochemical properties of the solid photocatalyst and are thus regarded as intrinsic factors. By comparison, θ_s in general building on the Langmuir-Hinshelwood kinetics is formulated as,[182]

$$\theta_s = \frac{K_{ad}C_{reac.}}{1 + K_{ad}C_{reac.}} \tag{5.4}$$

wherein K_{ad} and $C_{reac.}$ are the adsorption equilibrium constant and the concentration of the reactant yet present in the solution, respectively. Evidently, θ_s depends otherwise on the reaction condition and is thus regarded as an extrinsic factor.

Significantly, the independence of r_{ph} from C_{dye} in this study suggests that the present photocatalysis driven by the β-SnWO$_4$ spikecubes falls into an extreme of θ_s approximated to unity. Such scenario arises most likely from the moderate adsorption capacity of the β-SnWO$_4$ spikecubes due presumably to modest surface area (Table 5.1) that is extracted from the BET analysis (Fig. 5.4). In addition, little amount of the β-SnWO$_4$ spikecubes employed in the measurement more or less takes the responsibility likewise. In such context, intrinsic photocatalytic activity of the β-SnWO$_4$

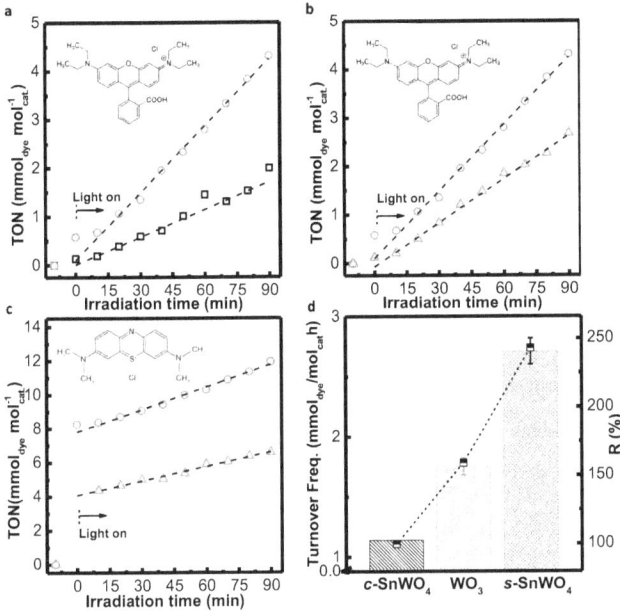

Figure 5. 3. Photocatalytic degradation of organic dyes under monochromatic light irradiation (wavelength: 366 nm) in the presence of either (a) β-SnWO$_4$ cubes (c-SnWO$_4$) or spikecubes (s-SnWO$_4$) for RhB degradation. Either β-SnWO$_4$ spikecubes or commercial WO$_3$ photocatalyst employed for (b) RhB and (c) MB degradation. Insets are the molecular structures of (a,b) RhB and (c) MB, respectively. (d) Photodegradation rate in terms of TOF (left axis) of β-SnWO$_4$ in cubic and spikecubic forms, and commercial WO$_3$ photocatalyst and the enhancement factor (right axis) derived via normalizing to the TOF of β-SnWO$_4$ microcubes.

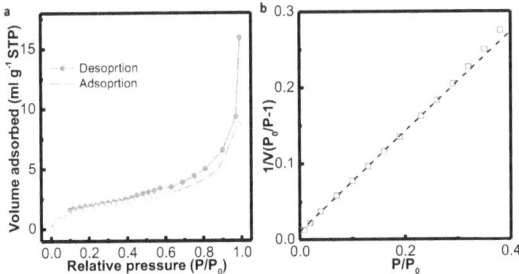

Figure 5. 4. (a) Adsorption isotherms at liquid nitrogen temperature (77 K) and (b) BET plot of the β-SnWO$_4$ spikecubes.

Table 5. 1. Geometric features and specific surface area of the β-SnWO$_4$ cubes, β-SnWO$_4$ spikecubes and commercial WO$_3$ photocatalyst.

Morphology	Size (μm)	Surface area (m²/g)
β-SnWO$_4$ cubes (Fig. 4.2c)	2-9	1.43 ± 0.10
β-SnWO$_4$ spikecubes (Fig. 4.2d)	2-9 (underlying cube) 0.7-2 (protrusive arm)	6.52 ± 0.91
Fine-particulate WO$_3$ (Fig. 5.7)	0.2	5.87 ± 0.61

spikecubes in the form of $\Phi_{ph}k_{red}C_s/k_r$ can be straightforwardly extracted from the measured photodegradation rate in terms of the turnover frequency (TOF).

$$TOF = \frac{TON}{Photoirradiation\ duration} \tag{5.5}$$

More significantly, the efficacy of the nanoarchitecture on the photocatalytic activity is highlighted by the superior photodegradation rate (Fig. 5.3a) of the β-SnWO$_4$ spikecubes to that of the β-SnWO$_4$ benchmark (Fig. 4.2a) otherwise in microcubic form (Fig. 4.2c). The enhancement exceeds 200% (Fig. 5.3d), which stems presumably from the open framework (Fig. 4.7) and multibranched structure (Fig. 4.2d) of the nanospike array effectively offering additional surface area to that of the underlying microcubes (Table 5.1 and Fig. 5.5), which in turn reinforces C_s of the β-SnWO$_4$ spikecubes. Particularly, the areal increment well remedies concomitant expense in Φ_{ph}, provided that the nanospikes unlikely harvests as much light as that by the microscale cubes due presumably to the nanoscale dimension merely comparable to the light penetration depth (Fig. 5.6a).[44]

To gain further insight into the photocatalytic properties of the β-SnWO$_4$ spikecubes, fine-particulate WO$_3$ with homogeneous particle size of ca. 200 nm (Fig. 5.7a), which is comparable to the nanospike size (Table 4.1) of the β-SnWO$_4$ spikecubes, is employed as an additional benchmark.[183] Such dimensional analogy gives rise to the surface area (Table 5.1) of the WO$_3$ benchmark approximating to that of the β-SnWO$_4$ spikecubes, which in turn suggests similar C_s. Moreover, this approximation and the analogous absorption coefficient of the WO$_3$ benchmark at wavelength of 366 nm to that of the β-SnWO$_4$ spikecubes (Fig. 5.7b) indicates a comparable Φ_{ph}. Surprisingly, the β-SnWO$_4$ spikecubes yet surpasses the WO$_3$ benchmark by more than 150% (Fig. 5.3d) in photocatalytically degrading either RhB (Fig. 5.3b) or MB (Fig. 5.3c). More importantly,

74

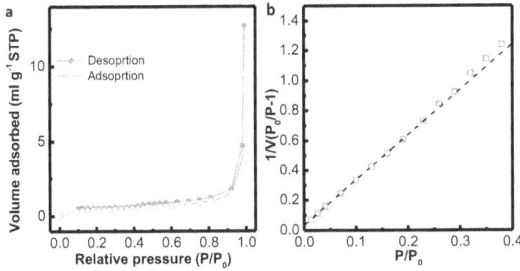

Figure 5. 5. (a) Adsorption isotherm at liquid nitrogen temperature (77 K) and (b) BET plot of the β-SnWO₄ microcube.

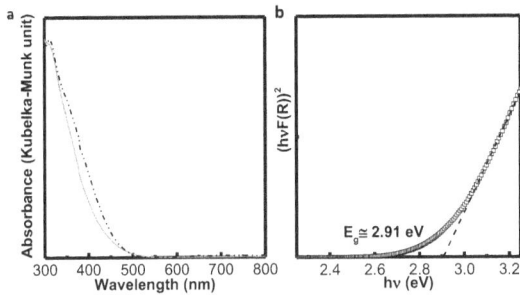

Figure 5. 6. (a) Kubelka-Munk transformed diffuse reflectance UV-VIS spectra of the β-SnWO₄ spikecube (red solid line) and the β-SnWO₄ microcube (block dash dot line). (b) Tauc plot (for direct interband transition) of the β-SnWO₄ microcube.

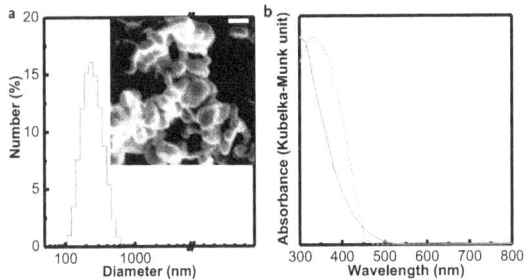

Figure 5. 7. (a) Size distribution histogram of commercial WO₃ benchmark with fine-particulate form (inset in (a); scale bar: 200 nm). (b) Kubelka-Munk transformed diffuse reflectance UV-VIS spectra of the β-SnWO₄ spikecube (red solid line) and fine-particulate WO₃ (green dash line).

although MB and RhB demonstrate significantly differential adsorption equilibria, wherein MB always shows stronger adsorption affinity, the photocatalytic activities of the WO_3 benchmark and the β-SnWO$_4$ spikecubes extracted respectively from two measurements are nearly identical. Such agreement substantiates the photodegradation rate measured in this study faithfully reflecting intrinsic photocatalytic activity of the materials.

In such context, the discrepancy in photocatalytic activity between the β-SnWO$_4$ spikecubes and the WO_3 benchmark stems indeed from the differentiation in k_{red}/k_r. Particularly, earlier reports by Abe and co-workers argues that the bottleneck in k_{red}/k_r of the WO_3 photocatalyst lying in poor reducing power of the photoelectrons due to the positive CB minimum (ca. +0.5 V vs. NHE).[183] On this account, oxygen (O) K-edge XANES is employed to study the unoccupied electronic states of the β-SnWO$_4$ spikecubes in order to elucidate superior k_{red}/k_r. Moreover, *ab initio* XANES computations are carried out in parallel to provide electronic perspectives on the measured spectral features.

5.1.1   Nanoarchitecture-mediated Band Structure Reformation

To meet this end, the rocksalt-like crystal structure (Fig. 5.8a) formulated by Jeitschko et al. for β-SnWO$_4$, wherein $[WO_4]^{2-}$ tetrahedra (Fig. 5.8c) form a face-centered cubic lattice with Sn^{2+} filling the octahedral sites – is adopted in the simulation of the O K-edge XANES in terms of the excellent agreement of measured XRD pattern (Fig. 4.2b) with that in this report.[147-149] Unlike the fairly regular $[WO_4]^{2-}$ tetrahedra with average W-O distances of ca. 1.75 Å, the octahedral coordination around Sn^{2+} is highly distorted due to electrostatic repulsion between the $5s^2$ lone pair of Sn^{2+} and the oxygen atoms (Fig. 5.8b). This in turn leads to two distinct Sn-O distances of 2.214 and 2.810 Å and consequently two different oxygen atoms, O(1) and O(2), in β-SnWO$_4$ (Fig. 5.8d). Particularly, O(1) is coordinated to one W and three Sn atoms in an approximately tetrahedral form. By contrast, O(2) has a coordination number of only two with the bonding to one W and one Sn atom, respectively. Given the intimacy between the electronic states and the local chemical bonding, simulated O K-edge XANES is further deconvoluted into the spectra of O(1) and O(2), respectively.

The numerical computation was carried out using FEFF implementation within the framework of the real space full multiple scattering (FMS) theory. All muffin-tin spheres were automatically overlapped by 30% (AFLOP card) to reduce the effects of potential discrepancies at the muffin tins and in particular to analogize the peculiar Sn^{2+} $5s^2$ lone pair that was characterized as spatially extended electronic orbital with the localized electron density directed toward the triangular face formed by the O(1) atoms of the $[SnO_6]$ coordination octahedron (Fig. 5.8c). The numerical computations were performed for different cluster sizes and the convergence was reached for cluster sizes of ca. 8.4 Å corresponding to 145[152] atoms around the photoabsorber O(1)[O(2)]. Numerical O K-edge were performed for different cluster sizes and the convergence was reached for cluster sizes of ca. 8.4 Å corresponding to 145[152] atoms around the photoabsorber O(1)[O(2)]. Numerical O K-edge XANE structures were calculated in the presence of an appropriately screened core hole according to the final-state rule. Theoretical features were convoluted with a Lorentzian-shape function to account for the core-hole lifetime ($\Gamma = 0.16$ eV) broadening. An additional broadening effect (0.3 eV) responsible for technical inaccuracies and lattice vibrations (Debye-Waller factor) of experimental aspect was further included during the calculation procedure. In addition, the projected density of states (pDOS) building on the *ab initio* FMS calculation for the photoabsorbing oxygen atom and all neighboring atoms (W and Sn) were also included.

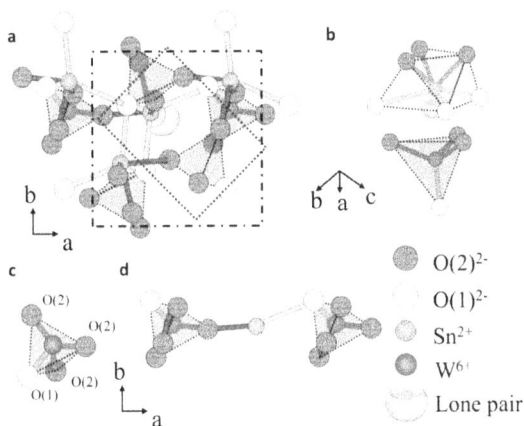

Figure 5. 8. (a) Projected stereodiagram of β-SnWO$_4$ with characteristic cubic crystal structure acts as theoretical standard model in numerical O K-edge XANES and W L_3-edge EXAFS calculations. Distorted [SnO$_6$] octahedron due to the electrostatic repulsion between characteristic $5s^2$ lone pair of Sn^{2+} and the lattice oxygen, which directs toward the triangular face formed by the O(1) atoms. (c) First oxygen coordinated shell around the W atom forms the fairly regular [WO$_4$] tetrahedron. (d) Diverse oxygen bridges (left: -O(2)-, right: -O(1)-) coordinate the [WO$_4$] tetrahedron to distorted [SnO$_6$] octahedron.

One takes into account that the well-known inaccuracies of energy-dependent exchange-correlation potentials during computation via a FEFF code typically lead to certain discrepancy between the calculated and the real Fermi levels on an energy scale of 1-2 eV.[184] This is particularly prominent for the systems containing f-electrons (e.g. W). On this account, an additional energy alignment of 1.49 eV is applied in the self-consistently obtained Fermi level using a FEFF code in this study, eventually giving rise to the excellent accord with the experimental XANES (Fig. 5.9a). Particularly, all oscillations including the strongest absorption peak (peak A) at 532 eV and an attached small shoulder (peak B', 533 eV) are well reproduced in the calculated XANES with a minor broadening effect. Such agreement renders the simulation highly informative, in turn favoring the subsequent discussions.

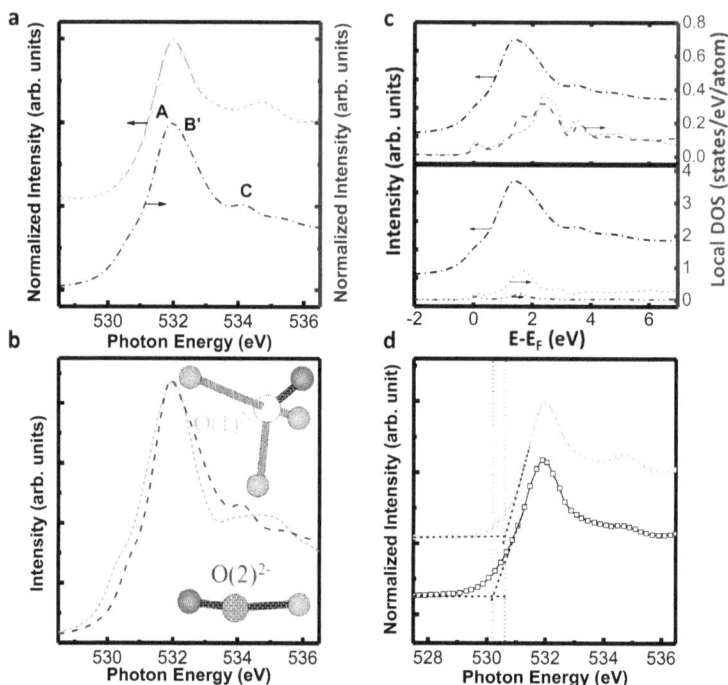

Figure 5. 9. (a) Experimental O K-edge XANE structure of the β-SnWO$_4$ spikecubes (red hollow sphere) along with the theoretical weight-averaged O K-edge XANE structure (blue dash dot line) based on the FMS calculation. (b) Numerical O K-edge XANE structures as a function of the nonequivalent O sites, O(1) (green dot line) and O(2) (purple dash line). (c) Projected DOS of (upper panel) non-metal O(1) (green dot line) and O(2) (purple dash line) relaxed $2p$ states, (lower panel), metal Sn $5s$ (blue dash dot dot line) and $5p$ (red dash line) states, and metal W $5d$ (grey dot line) state make up the weight-averaged O K-edge XANE structure (black dash dot line). (d) CB minimum revealed by the extrapolations of the leading edge in experimental O K-edge XANE structures of the β-SnWO$_4$ cubes (black hollow square) and spikecubes (red hollow sphere).

Numerical O K-edge XANE structures (Fig. 5.9b) for two nonequivalent O sites, O(1) and O(2), evidently reveal that the peak position of the main absorption (peak A) in weight-averaged (one O(1) and three O(2)) O K-edge XANE structure (Fig. 5.9a) is dictated by O(2) that is coordinated to one Sn and one W atom with respective bond lengths of 2.21 and 1.75 Å with an angle of 162.2° (lower inset in Fig. 5.9b).[147] In contrast, the absorption edge is otherwise determined by O(1) with higher coordination number (three Sn and one W) forming an off-centroid tetrahedron owing to the differences in the bond length between Sn-O(1) and W-O(1) and the closely neighbored Sn^{2+} $5s^2$ lone pair (Fig. 5.8b and upper inset in Fig. 5.9b).

Moreover, local DOS of the constituent elements making up the β-$SnWO_4$ spikecubes further reveal an electronic insight into these spectral oscillations (Fig. 5.9c). Significantly, the absorption edge in the numerical weight-averaged O K-edge XANE structure (on energy scale relative to the Fermi level (E-E_F)) in the first 2 eV above E_F builds on the majority vacant Sn $5p$ states and the minority unoccupied Sn $5s$, O(1) and O(2) $2p$ states. Given the higher coordination number of O(1) to Sn atoms (upper inset in Fig. 5.9c) and the closely neighbored Sn^{2+} $5s^2$ lone pair to O(1) (Fig. 5.8b), the strong hybridization between O(1) $2p$ and Sn $5s/p$ states accounts for the dominance of O(1) over the absorption edge in the weighted average structure that is simulated on the premise of the O K-edge excitation. By contrast, the major absorption (peak A) in weight-averaged structure consists of equivalently majority vacant Sn $5p$ and W $5d$ states, fractional empty O(2) $2p$ orbitals and the minority O(1) $2p$ and Sn $5s$ states. Given the characteristic coordination of O(2) to one W and one Sn atom (lower inset in Fig. 5.9c), the hybridization among O(2) $2p$, W $5d$ and Sn $5p$ states is responsible for the predominance of O(2) over the major absorption (peak A) in the numerical weight-averaged O K-edge XANE structures. By comparison, the featured oscillations beyond the major absorption (peak A) including shoulder B' and peak C are ascribed to the hybridization between the majority W $5d$ state, the minority Sn $5p$, O(1) and O(2) $2p$ states.

Specifically, the element- and orbital-resolved pDOS distributions derived herein from a multiple-scattering formalism inclusive of the relaxation effect to describe unoccupied conduction band characters of the β-$SnWO_4$ spikecube are in excellent agreement with earlier reports that relate to the ground-state calculation within the framework of the first principle density functional theory (DFT).[148] Particularly, the four-fold-coordinated [WO_4] tetrahedron in β-$SnWO_4$ undergoes a minor crystal field effect in comparison with the six-fold-coordinated [WO_6] octahedron in WO_3 that is the parental origin of bimetallic tungstates. This in turn centralizes the energetic distribution of the W $5d$ orbital that makes up CB of most filial tungstates upon the hybridization with O $2p$ states via antibonding interactions.[148,185-187] In consequence, the lowest unoccupied states in CB are negatively shifted (on scale of V vs. NHE), boosting the reducing power of the photoelectrons at the expense of photoabsorption efficiency due to the concomitant increment in E_g.[148,185] Such toll is low in β-SnWO$_4$ due to the presence of Sn^{2+}, rendering additional $5p$ and $5s$ characters to CB and VB that otherwise consists of the O non-bonding $2p$ states.[148] Particularly, the antibonding interaction between Sn $5s$ and O $2p$ orbitals significantly shifts the highest occupied levels in VB toward unoccupied CB, which effectively quenches the increment in E_g.[148,185-187] Such antibonding interaction is stabilized via a second-order-Jahn-Teller-distortion-mediated hybridization with empty Sn $5p$ states that make up the minimum of unoccupied CB. Such interaction suggests the expense of unoccupied conduction states, giving rise to the CB minimum at a more negative energy level (on scale of V vs. NHE), which further reinforces the reducing power of the photoelectrons of β-SnWO$_4$.

Evidently, the superior k_{red}/k_r of the β-SnWO$_4$ spikecubes to that of the WO$_3$ benchmark originates mostly from a reframed band structure that is dictated primarily by the crystallographic coordination, and moreover, mediated concurrently by the second-order Jahn-Teller (SOJT) distortion. More importantly, the comparison of unoccupied CB structure manifested in the O K-edge XANES between the β-SnWO$_4$ spikecubes and microcubic β-SnWO$_4$ benchmark (Fig. 5.9d) obviously punctuates the important role of the SOFT effect in the band structure reformation, provided that the β-SnWO$_4$ spikecubes and microcubes crystallize into the same structure (Fig. 4.2a,b). Particularly, the microcubic β-SnWO$_4$ benchmark exhibits a more dispersive distribution of unoccupied conduction states than that of the β-SnWO$_4$ spikecubes, as evidenced by the smooth leading edge in the O K-edge XANE structure (Fig. 5.9d). This in turn suggests that the lowest unoccupied states in CB is shifted toward the band gap, which is derived from the extrapolation of the leading edge in the O K-edge XANE structure. In other words, the CB minimum of the microcubic β-SnWO$_4$ benchmark is at a more positive energy level (on scale of V vs. NHE) with respect to that of the β-SnWO$_4$ spikecubes, which in turn results in a smaller E_g of 2.91($\pm$0.03) eV (Fig. 5.1b), and moreover, suggests a minor SOFT distortion in the β-SnWO$_4$ microcubes. To corroborate this tentative argumentation, W L_3-edge XAFS is additionally performed to uncover the atomic coordination of the β-SnWO$_4$ spikecubes and microcubes (Fig. 5.10 and Fig. 5.11).

At first sight, excellent agreement of the spectral profile and energy position of the white line in the W L_3-edge XAFS of the β-SnWO$_4$ spikecubes with that of the β-SnWO$_4$ microcubes (inset in Fig. 5.10) suggests that both adopt virtually analogous coordination framework despite of the pronounced surface nanoengineering (Fig. 4.2c,d). In contrast, the subtle change of the white line intensity may be attributed to one or more of the following effects: i) the charge transfer of the surface capping molecules, ii) an alternative bonding geometry around the W photoabsorber, and iii) the presence of a non-stoichiometric phase in either the β-SnWO$_4$ spikecubes or the microcubes.

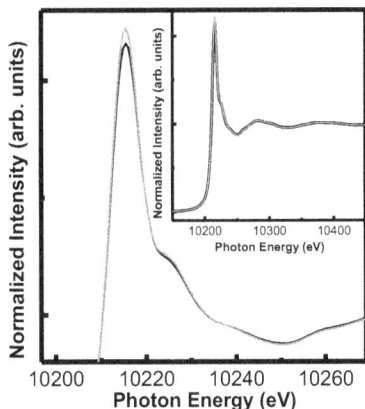

Figure 5. 10. Experimental W L_3-edge XAFS (inset) and XANES collected for spikecubic (red line) and cubic (black line) β-SnWO₄.

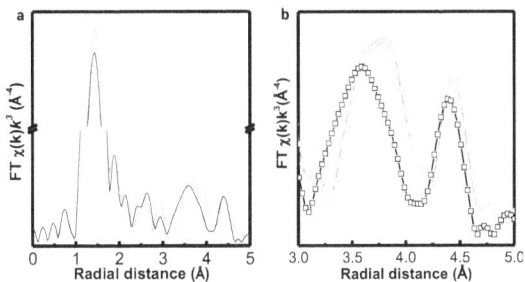

Figure 5. 11. Non-phase-corrected Fourier-transformed (FT) W L_3-edge EXAF structures $\chi(k)k^3$ for the β-SnWO₄ spikecubes (solid red line in (a) and hollow red sphere in (b)) and microcubic β-SnWO₄ benchmark (solid black line in (a) and hollow black square in (b)) as a function of radial distance of (a) 0-5 and (b) 3-5 Å with respect to photoabsorbing W site.

Given the microcubic and spikecubic β-SnWO₄ prepared in the context of the polyol synthesis, the crystal surface is mostly passivated by the DEG solvent via substituting near-surface oxygen atoms of β-SnWO₄ for those of DEG molecules.[143] Such homoatomic replacement suggests the absence of a charge transfer effect, whereas the steric hindrance of the concomitant ethylene group of DEG molecules otherwise reinforces static disorder at the crystal surface.[188] This in turn reduces the white line intensity, which turns out to be more significant with growing substitution. On such basis, the β-SnWO₄ spikecubes are expected to show minor white line intensity in the W L_3-edge XANES than that of the microcubic β-SnWO₄ benchmark in view of the nearly 5-fold higher specific surface

81

area (Table 5.1). Nevertheless, the experimental disagreement absolutely rules out such account of reduced white line intensity in the present contribution (Fig. 5.10).

Hence, Fourier transforms (FT) of the W L_3-edge EXAF structures with k^3 weighting ($\chi(k)k^3$) are subsequently performed to study the local coordination around the W photoabsorber (Fig. 5.11). The first FT maximum at a radial distance of 1.4 Å originates from the interference of photoabsorbing W atom with four nearest O scatters in the first coordination shell, forming four W-O bonds in tetrahedral [WO$_4$] form (Fig. 5.11a). Significantly, excellent agreement in the peak symmetry and radial distance of the first-shell peak suggests equivalently regular [WO$_4$] tetrahedra in both the β-SnWO$_4$ spikecubes and the microcubes. In contrast, a surprisingly pronounced deviation is present in the second FT maximum at the radial distance of 3.0-4.2 Å, which stems from the interference of the [WO$_4$] tetrahedra with the nearest Sn scatters (Fig. 5.11b). Moreover, the second-shell peak manifests either the bimodal or the somewhat asymmetric hump-like shape for the β-SnWO$_4$ spikecubes and the microcubes, respectively. Such profile discrepancy suggests the displacement of the Sn scatters with respect to the W absorbers, and moreover, evidences different local atomic structures of the β-SnWO$_4$ spikecubes and the microcubic β-SnWO$_4$ benchmark. Particularly, such structural adaption is in good accord with earlier studies on β-SnWO$_4$, which suggest that the relative positions of the rigid tetrahedral [WO$_4$] building units with respect to the nearest octahedral [SnO$_6$] entities can be readily modulated due to weak interactions in between.[189,190]

To quantitatively specify the local structure details including the coordination number (N), the bond length (R_{bond}) and the Debye-Waller factor (σ^2) of the W photoabsorbing site, theoretical fitting was performed on the real part of the FT W L_3-edge EXAFS $\chi(k)k^3$ (Table 5.2 and Fig. 5.12). In these fitting routines, the characteristic cubic structure of β-SnWO$_4$ (Fig. 5.8a) was likewise adopted.[147-149] Fits were performed in the R_{bond}-space for the first FT maximum via modulating R_{bond}, the bond length distribution width (σ) and NS_0^2, wherein $N = 4$ was employed that inferred from four O atoms in the first coordination shell around W, forming tetrahedral [WO$_4$] building units in β-SnWO$_4$ (Fig. 5.8c).[147-149] In contrast, the amplitude reduction factor (S_0^2) was evaluated graphically from scaling the average of several numerical fits to that of experimental scans, which is in general between 0.7 and 1.0. In this way, $S_0^2 = 0.72(3)$ was derived and then fixed in the subsequent fits for the second and third coordination shells that are characterized as the FT maxima at the radial distance of 3-5 Å in the FT W L_3-edge EXAFS $\chi(k)k^3$ (Fig. 5.11b).

Noteworthily, all the numerical fits trace the experimental scans in the real part of the FT W L_3-edge EXAFS $\chi(k)k^3$ fairly well – particularly in the cubic case – over the entire R_{bond} range (Fig.

5.12). Such excellent agreement suggests that the β-SnWO$_4$ spikecubes and the microcubes substantially adopt a cubic framework of model β-SnWO$_4$ (Fig. 5.8a), as indicated already by the XRD analysis (Fig. 4.2a,b) and numerical computation of the O K-edge XANES (Fig. 5.9). Four O atoms in the first coordination shell form the W-O bonds with an exclusive distance of 1.780 Å, erecting regular [WO$_4$] tetrahedra in both cubic and spikecubic β-SnWO$_4$ (Table 5.2).

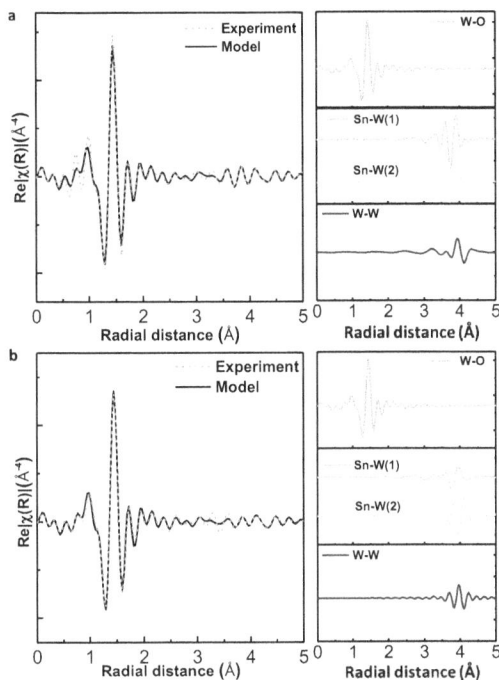

Figure 5. 12. Real part of W L_3-edge EXAF structures $\chi(k)k^3$ for (a) the β-SnWO$_4$ spikecube and (b) microcubic β-SnWO$_4$ benchmark. (a,b) Right panels demonstrated the individual contributions of distinct coordination shells as a function of R_{bond}.

The minor deviations of the W-O bond length from those (1.746 and 1.764 Å) reported for the model β-SnWO$_4$ are much smaller than the resolution limit of the EXAFS technique,[147] which in other words suggests the reliability of the derived structure metrics along with the uncertainty (in parenthesis) summarized in Table 5.2. Moreover, the reasonable σ^2 and goodness-of-fits (R_f) further reinforce the accuracy, albeit the uncertainty otherwise becomes significant when the fits are performed on the coordination shells beyond the first one. This is a common issue for the EXAFS technique and ascribed in the present study to one or more of the following factors: i) systematic

83

errors of the EXAFS fitting procedure, wherein the background subtraction more-or-less results in damped XAFS oscillations,[126,119,188] ii) the mild thermal conditions employed in the polyol syntheses of the β-SnWO$_4$ spikecubes and the microcubes, leading to mediocre crystallinity that in turn attenuates the intensity of the EXAFS fluctuations,[143] and iii) the presence of the defects and the disorders in the materials, which otherwise strengthen the uncertainty of R_{bond} and σ^2.[188] On this account, the focus in this contribution is preferentially given to the nanoarchitecture-mediated local structure transition in lieu of absolutely quantifying the local structure parameters.

Table 5. 2. Structure metrics derived from the numerical fitting of the W L_3-Edge EXAFS of spikecubic and cubic β-SnWO$_4$.[ab]

Absorber-backscatter pair	N	R_{bond} (Å)	σ^2 (Å^2)
β-SnWO$_4$ spikecube			
W-O	4	1.780(1)	0.0010(2)
W-Sn(1)	2.1(9)	4.050(74)	0.0067(8)
W-Sn(2)	1.6(7)	4.240(66)	0.0043(59)
W-W	0.6(3)	4.540(24)	0.0010(59)
β-SnWO$_4$ microcube			
W-O	4	1.780(12)	0.0010(2)
W-Sn(1)	4.0(1.7)	4.080(35)	0.0147(40)
W-Sn(2)	0.9(4)	4.250(25)	0.0029(20)
W-W	0.5(4)	4.540(31)	0.0010(59)

Notation: Effective coordination number (N), interatomic distance (R_{bond}) and Debye-Waller parameters (σ^2).

Typical β-SnWO$_4$ with cubic structure is used as the model for EXAFS calculations.[147-149]

Statistical error for the associated model parameters: $N \pm 20\%$; $R_{bond} \pm 1\%$; $\sigma^2 \pm 20\%$.

[a]Fitting range: k = 3.5-14.5 Å^{-1}, R_{bond} = 1.0-5.0 Å. Values in parentheses are uncertainties in the least significant digit estimated for the corresponding structure metrics (associated with fits).

[b]Fit quality index R_f (R-factor = $\sum$(data-fit)^2/$\sum$data^2) = 0.01. Values of other EXAFS model parameters not shown above are either fixed or fitted to a common value over all samples as follows: S_0^2 = 0.72(0.03) (fixed amplitude reduction factor based on first-shell fitting to spikecubic β-SnWO$_4$ with improved crystallinity); ΔE_0 = −0.059 eV (fitted energy shift parameter).

Numerical fits to disclose such structural transformation that is of relevance to the displacements of the Sn atoms in the second coordination shell (Fig. 5.11b) begin with the microcubic β-SnWO$_4$ benchmark. In the β-SnWO$_4$ microcubes, most Sn scatters (Sn(1)) are distributed at a radial distance of 4.080(35) Å with respect to the W absorber via either W-O(1)-Sn or W-O(2)-Sn bridge (Fig. 5.8b). Moreover, these are the majority atoms in the second coordination shell around the W site, which amount to 4.0(1.7), as suggested by the effective N (Table 5.2).

Additional Sn scatters (Sn(2)) are otherwise dispersed at a longer radial distance of 4.250(25) Å with respect to the W absorber. Such distribution alternatively results from electrostatic repulsion between the non-bonding $5s^2$ lone pair and the lattice oxygens that establish octahedral [SnO_6] building units of β-$SnWO_4$, leading to the off-centroid displacement of the Sn atom in the direction away from the W site (Fig. 5.8c). These are the minority atoms in the second coordination shell around the W site, amounting only to 0.9(4), as indicated by the corresponding N (Table 5.2). The β-$SnWO_4$ microcubes reach good agreement with the reference β-$SnWO_4$ model employed in the fitting routines on the relative quantities of two distinct Sn atoms along with the overall population in the second coordination shell around the W photoabsorber.[147-149] Given the damped EXAFS fluctuations of the microcubic β-$SnWO_4$ benckmark presumably due to the mediocre crystallinity (Fig. 4.2a), the underestimation of the amount of Sn is highly likely present in numerical fits, which is responsible for minor discrepancy.

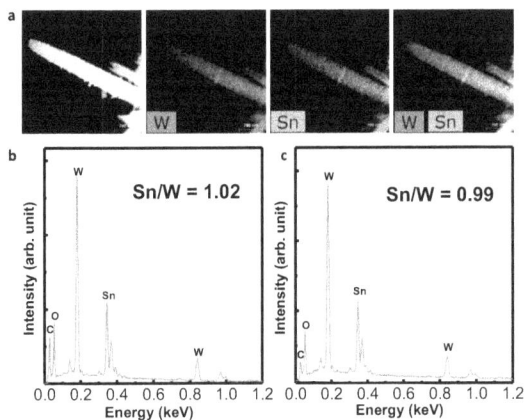

Figure 5. 13. (a) HAADF-STEM images (leftest panel) and corresponding EDXS elemental mapping (right three panels) demonstrated the atomic distribution of heavy W (red) and Sn (green) elements over the surface nanospike site. (b,c) Global composition analyses via the EDXS technique of a β-$SnWO_4$ spikecube (b) and a microcube (c).

Numerical fits are then performed for the β-$SnWO_4$ spikecubes likewise on the second coordination shell. At first sight, two distinct Sn scatters in the second coordination shell around the W site are successfully deconvoluted with characteristic W-Sn interatomic distances of 4.050(74) and 4.240(66) Å, respectively. Surprisingly, the discrepancy in either W-Sn distance between microcubic and spikecubic β-$SnWO_4$ is much smaller than the resolution limit of the EXAFS technique, leading to the Sn scatters of spikecubic β-SnWO$_4$ to be likewise categorized into the same groups, viz. Sn(1) and Sn(2), respectively.[147-149] Such conformity in W-Sn radial distances suggests the nanoarchitecture-mediated textural adaption manifested exclusively in the respective quantities of distinct Sn scatters in the second coordination shell, which are characterized by the effective N (Table 5.2).

Noteworthily, the population of the overall Sn scatters of a β-SnWO$_4$ spikecube is evidently lower with respect to those of the microcubic benckmark and the reference model.[147-149] This is unlikely an underestimation by numerical fits, provided that the crystallinity of the β-SnWO$_4$ spikecubes surpasses that of microcubes (Fig. 4.2a,b), which otherwise reinforces the intensity of the EXAFS oscillations. In contrast, it is more likely ascribed to the nanoengineering introducing more undercoordinated atoms to the surface of the β-SnWO$_4$ spikecube. Particularly, the undercoordination preferentially is present at the O-sites in the second coordination shell in terms of the high stoichiometry with respect to Sn and W, as suggested by the composition analysis using the EDXS technique (Fig. 5.13). Particularly, local characterization suggests a homogeneous distribution of equivalent Sn and W over the surface nanospike sites, reaching good agreement with the global quantification of the β-SnWO$_4$ spikecubes and microcubic β-SnWO$_4$ benckmark on the stoichiometric ratio of Sn to W approximating to unity.

This argumentation is alternatively reinforced by numerical simulation of O K-edge XANE structure (Fig. 5.9a and Fig. 5.14). Particularly, the absorption oscillation (peak C) far beyond the leading edge is dictated primarily by O(2) characterized by a lower coordination number (one Sn and one W) in view of the excellent accord in peak position (Fig. 5.14a). This is ascribed mostly to the weight-averaged O K-edge XANE structure derived primarily from weighting the individual contributions of the nonequivalent O(1) and O(2) atoms on the stoichiometric basis, wherein one O(1) and three O(2) atoms are present in the unit cell of the β-SnWO$_4$ model (Fig. 5.8a), followed by averaging out, resulting in the predominance of O(2) over peak C in the end.[147-149] In contrast, the absorption feature C in the experimental O K-edge XANE structure of the β-SnWO$_4$ spikecubes is at an energy position more consonant with that in the numerical O K-edge XANE structure derived at the O(1) site (Fig. 5.14b). Such energetic conformity clearly indicates that the surface undercoordination is presumably of relevance to the oxygen vacancies present preferentially at the O(2) sites. More importantly, these atomic defects represent more empty space available in the second coordination shell for the majority Sn scatters, which is responsible for the displacement manifested in the perturbation of the effect N of the β-SnWO$_4$ spikecubes (Table 5.2). Most significantly, the presence of those vacancies and disorders indicates in other words highly distorted crystal structure of the β-SnWO₄ spikecubes. Eventually, this corroborates with the former argument that the nanoarchitecture-triggered structural distortion reframes the band structure of the β-SnWO₄ spikecube via a SOFT effect, which accounts for the superior photocatalytic activity to those of the fine-particulate WO₃ and the microcubic β-SnWO₄ benchmark, respectively (Fig. 5.3b).

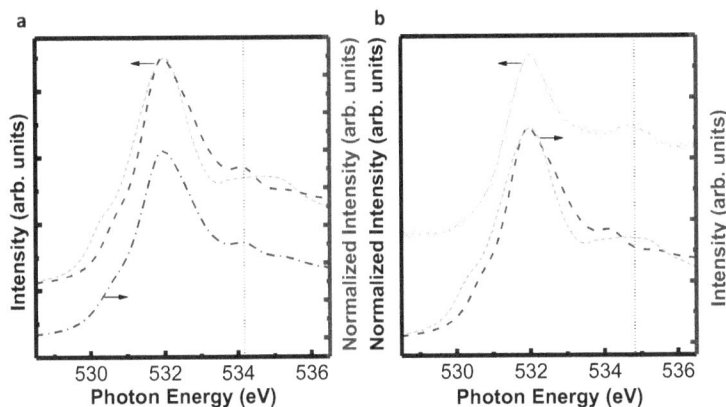

Figure 5. 14. Individual contributions of (green dot line) O(1) and (purple dash line) O(2) atoms to the (a) weight-averaged O K-edge XANE structure (blue dash dot line) and (b) the experimental O K-edge XANES of the β-SnWO₄ spikecube (red hollow sphere).

In summary, surface nanoarchitecture not only endows the β-SnWO₄ spikecubes with superior adsorption capacity due to an enhanced specific surface area. Moreover, a large number of undercoordinated atoms are introduced to surface nanospike array, which serve as the majority reaction sites for the photocatalytic decomposition of dye molecules. Particularly, reinforced SOJT distortion preferentially at these non-equilibrium sites significantly reframes the surface band structure of the β-SnWO₄ spikecubes. Altogether, the synergistic effect of fabricating β-SnWO₄ with a stereoactive lone pair into a multibranch-shaped hierarchical nanoarchitecture accounts for superior photocatalytic activity manifested in the dye degradation metrics.

5.1.2 Nanoarchitecture-mediated Broadband Photoabsorption

The photocatalytic activity of another topical artifact, viz. the Au@Nb@H$_x$K$_{1-x}$NbO₃ nanopeapods, is likewise evaluated via the dye photodegradation metrics with some modifications, as discussed below. The reaction was carried out in a Pyrex reaction cell containing a suspension of powdery photocatalyst (50 mg) under investigation in 100 mL of aqueous dye solution with continuous agitation using a magnetic stirrer. Before illumination, the suspensions were magnetically stirred in the dark for 1 h to ensure the establishment of an adsorption/desorption equilibrium of the dyes on the photocatalyst surface. During photoirradiation, 2 ml of aliquot were sampled from the reaction reservoir every 30 minutes. Centrifugation was applied to separate the photocatalyst from the dye solution that was then analyzed via UV-VIS spectrophotometry. The concentration evolution of the dye solution was derived from the absorbance variation registered at the signature wavelengths of 665 and 554 nm for MB and RhB, respectively (Fig. 5.15a). Noteworthily, a diffusive simulated AM 1.5 G sunlight at a fluence of 100 mW cm^{-2} was utilized herein as the illumination source.

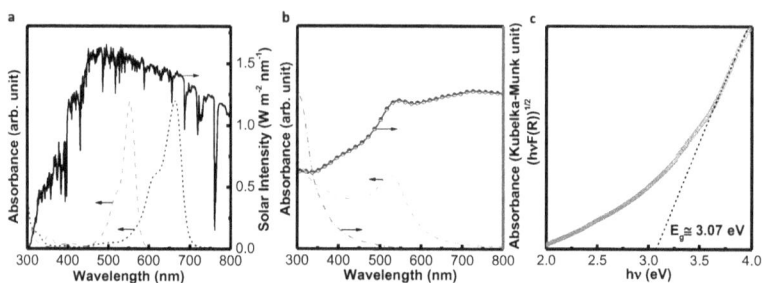

Figure 5. 15. (a) UV-VIS absorption spectra (left axis) of aqueous RhB (brown dash line) and MB (royal blue dot line) solutions. (right axis) AM 1.5 G solar spectrum (black solid line) is plotted alongside for comparison. (b) Kubelka-Munk transformed diffuse reflectance spectra (right axis) of the Au@Nb@H$_x$K$_{1-x}$NbO$_3$ nanopeapods (blue half-filled circles) and hollow H$_x$K$_{1-x}$NbO$_3$ nanoscrolls (green dash dot line). UV-VIS absorption spectrum (left axis) of sub-10 nm Au nanoparticles (orange dashed line) is plotted alongside for comparison. (c) Tauc plot (for indirect interband transition) of hollow H$_x$K$_{1-x}$NbO$_3$ nanoscrolls.

Such scheme originates from the characteristic photoabsorption property of the Au@Nb@H$_x$K$_{1-x}$NbO$_3$ nanopeapods toward not only UV but also visible up to near-infrared (NIR) light (Fig. 5.15b). This broadband light harvesting ability is ascribed to the constituents, and more importantly, the biomimicry of the peapod design, as elucidated below. The outmost semiconducting H$_x$K$_{1-x}$NbO$_3$ nanosheaths of the Au@Nb@H$_x$K$_{1-x}$NbO$_3$ nanopeapods exhibit a band gap of 3.07 eV (Fig. 5.15c) that is derived from the Kubelka-Munk transformed diffuse reflection spectrum (Fig. 5.15b) using the Tauc law (Eqn. 3.13). This accounts primarily for the photoabsorption of the Au@Nb@H$_x$K$_{1-x}$NbO$_3$ nanopeapods toward the UV light (Fig. 5.15b). In contrast, the inmost core-shell Au@Nb bimetallics are responsible for the photoabsorption of the Au@Nb@H$_x$K$_{1-x}$NbO$_3$ nanopeapods toward the visible light. In the present contribution, the coinage metal Au characterized by the sub-10 nm nanocrystal size (Fig. 4.14a) harvests visible photons preferentially with the wavelength of ca. 525 nm through an excitation of localized surface plasmon resonance (LSPR), as manifested in the UV-VIS absorption spectrum (Fig. 5.15b).

Virtually, LSPR is theoretically feasible in any metal, alloy or semiconductor with a large negative real dielectric constant and a small positive imaginary part of the dielectric constant (Fig. 5.16a,b).[191-193] Surprisingly, the refractory metal Nb satisfies such requisites (Fig. 5.16a,b) and earlier numerical calculations suggest that the resonant wavelength of LSPR specific to 10 nm sized Nb nanocrystal is approximately of 400 nm (Fig. 5.16c).[192-194] The core-shell form of the Au@Nb bimetallic in general leads to the LSPR of the Au core rapidly shielded by the Nb shell.[195] Such masking effect is more-or-less circumvented in this study due to the superior LSPR of Au to that of Nb (Fig. 5.16c), and more importantly, the formation of a very thin Nb shell of only few atomic layers thick (Fig. 4.17a).[194,195] This in turn gives rise to the coexistence of characteristic LSPRs of the Au core and the Nb shell in the Kubelka-Munk transformed diffuse reflection spectrum of the Au@Nb@$H_xK_{1-x}NbO_3$ nanopeapods, which exhibits a somewhat flatter nature at wavelengths of 400-550 nm (Fig. 5.15b). Certain red-shift of the resonant wavelength of two LSPRs of bimetallic Au@Nb is otherwise ascribed to the localized increase in the surrounding refractive index, provided that these sub-10 nm Au@Nb nanocrystals are brought into contact to tubular $H_xK_{1-x}NbO_3$ nanoscrolls to form the Au@Nb@$H_xK_{1-x}NbO_3$ nanopeapods (Fig. 4.13).[195,196]

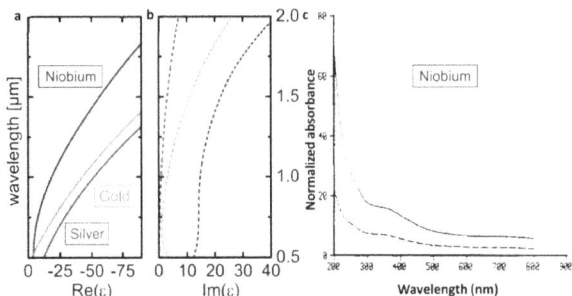

Figure 5. 16. (a) Real (solid lines) and (b) imaginary parts (dashed curves) of the dielectric functions of Au (green), Ag (purple) and Nb (blue), respectively. (c) Calculated absorption spectra of 10 nm sized spherical Nb nanoparticles dispersed either *in vacuo* (dash line) or in a dielectric medium (solid line) with the refractive index of 1.33 that is equivalent to that of water (*adapted from reference[192,194]*).

Evidently, the building units of the Au@Nb@$H_xK_{1-x}NbO_3$ nanopeapods account only for the photoabsorption toward UV and visible light with wavelengths up to 550 nm. In other words, the bathochromic absorption band in the red-NIR region (Fig. 5.15b) is attributed to the biomimicry of the peapod design, as elaborated below. Statistical analysis over hundreds of TEM images clearly reveals the narrow distribution of the particle size of bimetallic core-shell Au@Nb nanocrystals, which are characterized by an average diameter of 8.8 ± 1.3 nm (Fig. 5.17a). Particularly, most Au@Nb nanocrystals are in close proximity (Fig. 5.17c) with a highly uniform interparticle distance of ca. 2 nm (Fig. 5.17b).

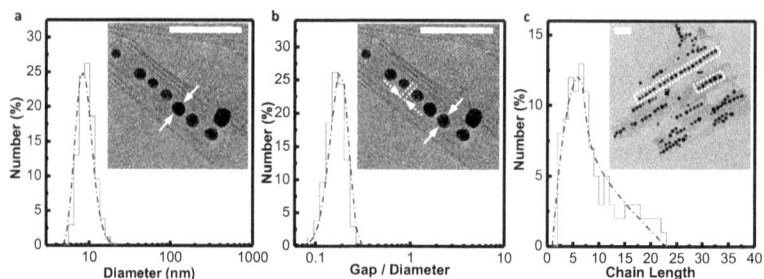

Figure 5. 17. Statistical analysis of (a) the particle size (solid arrays in the representative TEM image (scale bar: 50 nm) in the inset) of the core-shell Au@Nb bimetallic, (b) the interparticle distance (dash arrays in the representative TEM image (scale bar: 50 nm) in the inset) expressed in units of the Au@Nb particle size (solid arrays in the representative SEM image) and (c) the length of continuous yet discrete Au@Nb chain (dash frames in the representative TEM image (scale bar: 50 nm) in the inset) expressed in numbers of Au@Nb nanocrystals per chain over hundreds of TEM images. The statistics is made on the premise that the Au@Nb nanocrystals distributed in the peapod configuration with respect to the $H_xK_{1-x}NbO_3$ nanoscrolls.

   Given LSPR described as the resonant photon-induced collective oscillation of valence electrons at the metal surface, such nanometric resolution over sub-microscopic distance absolutely favors the near-field plasmon-plasmon coupling between these bimetallic entities.[197-200] Such electric near-field effect modulates the intrinsic LSPR of spherical Au@Nb nanocrystals in transverse mode that refers to the oscillatory direction of surface electrons of metal perpendicular to the chain axis (upper inset in Fig. 5.18).[196] Given the electrons of all the nanoparticles oscillating in the same phase, the restoring force of such electric dipole against the positive metallic nuclei increases due to the charge distribution of the neighbored metallic nanocrystals repelling the dipole. In consequence, the resonant wavelength is shifted to the blue end of the spectrum with respect to that of an isolated single nanocrystal. Moreover, the near-field coupling excites an addition LSPR in longitudinal mode that refers to the oscillatory direction of surface electrons of metal otherwise parallel to the chain axis (lower inset in Fig. 5.18).[196] In contrast to the dipolar interaction in transverse mode, electric dipole in this case is attracted by the surface charges of neighbored metallic nanocrystals, undergoing otherwise a minor restoring force. As a result, such longitudinal dipolar interplay endows the spherical Au@Nb nanocrystals with an extra LSPR at the red end of the spectrum, which accounts for the bathochromic absorption band distributed over nearly the entire red-NIR regime up to 800 nm in the Kubelka-Munk transformed diffuse reflection spectrum of the Au@Nb@$H_xK_{1-x}NbO_3$ nanopeapods (Fig. 5.15b).

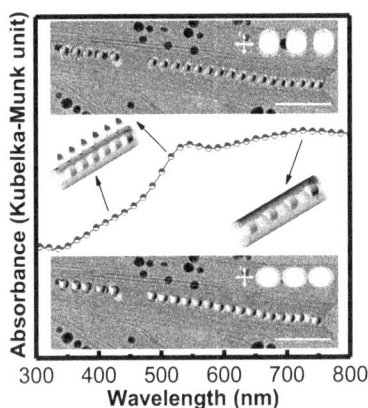

Figure 5. 18. Schematic illustrations of the near-field dipolar coupling between close yet discrete metallic nanocrystals modulates the resonant mode in the transverse direction (upper and left insets) and excites an additional resonant mode in the longitudinal direction (lower and right insets) with respect to the chain axis.

Insofar, the broadband photoabsorption of the Au@Nb@$H_xK_{1-x}NbO_3$ nanopeapods is explicitly ascribed to i) the interband electronic transition of semiconducting $H_xK_{1-x}NbO_3$ nanoscrolls via the absorption of the UV photons, the excitation of LSPRs ii) in transverse mode of bimetallic Au@Nb nanocrystals by the visible photons with wavelength up to 550 nm, and iii) in longitudinal mode by the red-NIR photons, respectively. On this account, additional UV-cutoff ($\lambda >$ 420 nm) and colored glass ($\lambda >$ 610 nm) filters were further coupled with the diffusive sunlight irradiation source in order to unambiguously correlate the photocatalytic dye degradation performance with the charge carriers photogenerated through diverse mechanisms that are highly wavelength-dependent (Fig. 5.19).

Evidently, the presence of bimetallic Au@Nb nanocrystals substantially accelerates the semiconducting $H_xK_{1-x}NbO_3$ nanoscrolls to photobleach the dye solution (Fig. 5.19d-i) under either full sunlight (Fig. 5.19a,d,g), integral VIS-NIR light (Fig. 5.19b,e,h) or red-NIR light (Fig. 5.19c,f,i)

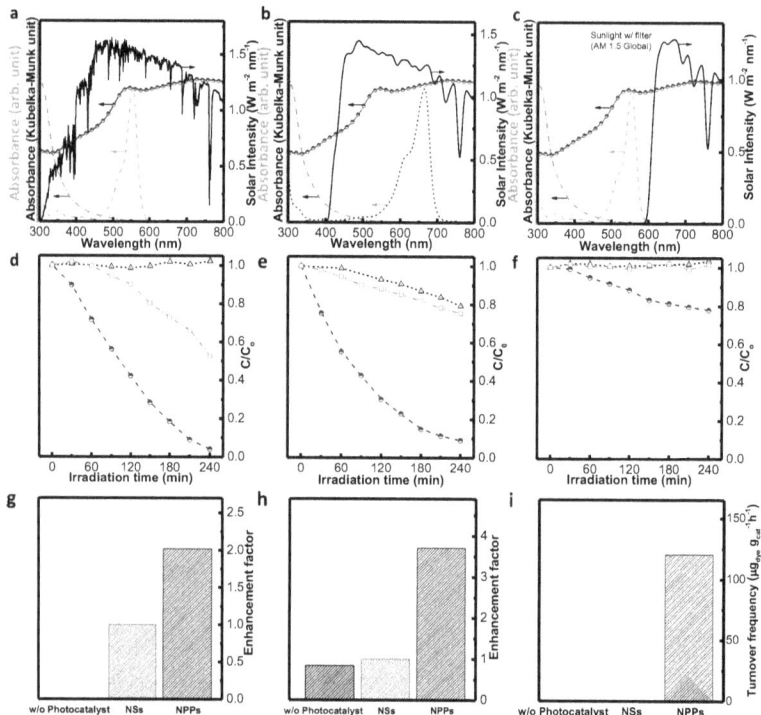

Figure 5. 19. (Right axis) Spectra of simulated sunlight (a) further equipped with either an UV cutoff (b) or a colored glass filter (c) that are employed as the irradiation sources in the photocatalytic dye degradation metrics, respectively. (Left axis) Kubelka-Munk transformed diffuse reflection spectrum of the $H_xK_{1-x}NbO_3$ nanoscrolls (green dash dot line) and the Au@Nb@$H_xK_{1-x}NbO_3$ nanopeapods (blue half-filled circles) employed as the photocatalysts. (Left axis) UV-VIS absorption spectra of MB (royal blue dot line) and RhB (brown dash line) utilized as the probing species (dashed line) are plotted alongside for reference. (d-f) Photocatalytic decoloring of RhB (d,f) and MB (e) in the absence (black hollow triangles) or in the presence of either the $H_xK_{1-x}NbO_3$ nanoscrolls (green hollow squares) or the Au@Nb@$H_xK_{1-x}NbO_3$ nanopeapods (blue half-filled circles). (g-i) Comparison of the wavelength-dependent photocatalytic activity between the $H_xK_{1-x}NbO_3$ nanoscrolls (green columns) and the Au@Nb@$H_xK_{1-x}NbO_3$ nanopeapods (blue columns) under illumination of full solar light (g), integral VIS-NIR light (h) and red-NIR light (i), respectively.

92

illumination, respectively. The best decoloring performance is derived in the case of the Au@Nb@$H_xK_{1-x}NbO_3$ nanopeapods under full solar irradiation (Fig. 5.19d). Significantly, only the minority charge carriers arose from the $H_xK_{1-x}NbO_3$ nanoscrolls via the UV excitation involve in this chemical conversion. In contrast, the inmost core-shell Au@Nb nanocrystals offer the majority carriers via both transverse and longitudinal LSPRs that account for the acceleration that is expressed in terms of the enhancement factor (Fig. 5.19g), which is derived from the normalization of the apparent turnover frequency (TOF) to that of the $H_xK_{1-x}NbO_3$ benchmark.

$$\text{TOF} = \frac{Weight\ of\ reacted\ molecules}{[Weight\ of\ the\ photocatalyst][Photoirradiation\ duration]} \qquad (5.6)$$

Particularly, the Au@Nb nanocrystals contribute those charge carrier via either directly populating energetic electrons to adjacent $H_xK_{1-x}NbO_3$ (direct electron transfer, DET) or resonantly transferring the plasmonic energy to $H_xK_{1-x}NbO_3$ to excite more e^-/h^+ pairs (resonant energy transfer, RET).[196,201-203] The minor spectral overlap between bimetallic Au@Nb nanocrystals and the $H_xK_{1-x}NbO_3$ nanoscrolls (Fig. 5.15b and Fig. 5.16c) excludes RET from the charge delivery mechanism. In the DET process, the electrons of plasmonic metal derive energy from the resonant photons, which then traverse the metal-semiconductor interface – the well-known Schottky junction – and migrate to CB of $H_xK_{1-x}NbO_3$. On such basis, the internal quantum transmission probability (η_i) of this electronic injection follows the modified Fowler equation,[204,205]

$$\eta_i \propto \frac{(h\nu - q\Phi_B)^2}{h\nu} \qquad (5.7)$$

wherein $h\nu$ and $q\Phi_b$ are the incident photon energy and energetic barrier height at the Schottky interface, respectively. $q\Phi_b$ is established by electronic alignment between the Fermi level of plasmonic Au@Nb bimetallic and the flatband potential (V_{fb}) of the $H_xK_{1-x}NbO_3$ semiconductor. The work function of the core-shell Au@Nb nanocrystal is located in a spectrum with Au (5.1 eV) and Nb (4.3 eV) standing at the ends, approximating most likely to the Au boundary in view of the predominance of Au in the composition stoichiometry over that of Nb, as validated by the quantitative EDXS analysis (Fig. 5.20).[206] V_{fb} of the $H_xK_{1-x}NbO_3$ semiconductor is otherwise estimated using the equation formulated by Scaife et al. for oxides that do not contain metal cations with partially filled d orbital,[207]

$$V_{fb}(vs.\ NHE\ ) = 2.94 - E_g \qquad (5.8)$$

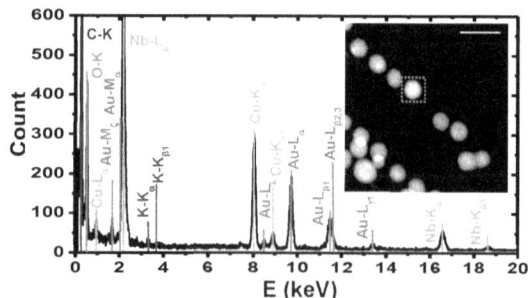

Figure 5. 20. (a) EDX spectrum (performed at the framed region in the HAADF-STEM image (inset, scale bar: 20 nm) of the Au@Nb@$H_xK_{1-x}NbO_3$ nanopeapods) ascribes an averaged composition stoichiometry of $Au_{0.67}Nb_{0.33}$ to the core-shell Au@Nb bimetallics. The quantification is carried out on the premise that the chemical composition of the outmost niobate nanoscrolls is $H_{0.97}K_{0.03}NbO_3$ (Fig. 4.11), which is subtracted from the analysis.

wherein E_g is the bandgap of the semiconductor. On such basis, V_{fb} of -0.13 V (vs. NHE) for the $H_xK_{1-x}NbO_3$ semiconductor and a Schottky barrier height of approximately 0.7 eV between plasmonic Au@Nb bimetallics and the $H_xK_{1-x}NbO_3$ nanoscrolls are concluded (Fig. 5.21).

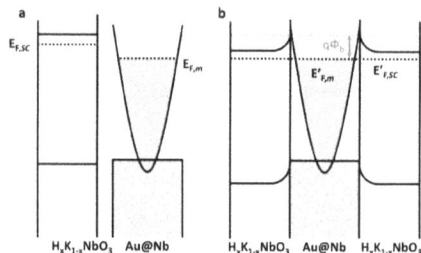

Figure 5. 21. (a) Energy diagrams of the isolated $H_xK_{1-x}NbO_3$ semiconductor and the core-shell Au@Nb bimetallic. (b) Formation of the Schottky junction when the Au@Nb nanocrystal is enclosed by the $H_xK_{1-x}NbO_3$ nanoscroll.

Surprisingly, this energy barrier is nearly one quadrant of E_g of the $H_xK_{1-x}NbO_3$ semiconductor, which is the key to turn visible and NIR photons that are invisible to the $H_xK_{1-x}NbO_3$ nanoscolls available to the Au@Nb@$H_xK_{1-x}NbO_3$ nanopeapods (Fig. 5.18). More importantly, those photons effectively energize the electrons of plasmonic Au@Nb bimetallics via the excitation of LSPRs in transverse and longitudinal modes, respectively. This is in favor of those hot electrons overcoming the Schottky barrier to enter the $H_xK_{1-x}NbO_3$ nanoscrolls and eventually participating in chemical transformation of the dye molecules. The steady and efficient photobleaching courses in the

94

presence of the Au@Nb@H$_x$K$_{1-x}$NbO$_3$ nanopeapods under photoirradiation of integral VIS-NIR (Fig. 5.19e,h) and red-NIR light (Fig. 5.19f,i) are the prima facie evidences.

5.2 Photoelectrochemical Water Splitting Metrics

Nevertheless, the argumentation regarding the visible- and NIR-light-active photocatalytic properties of the Au@Nb@H$_x$K$_{1-x}$NbO$_3$ nanopeapods, which builds on the dye photodegradation metrics, fails to be solid and persuasive. The uncertainty originates primarily from the MB and RhB dyes that are highly colorful and exhibit strong absorption bands in the red and visible wavelength regimes (Fig. 5.19b,c). In other words, an explicit quantification of the photocatalytic performance in terms of either apparent or "true" quantum efficiencies hardly meets, provided that the determination of the extent to which the incident photons are absorbed by either the dye molecule or the solid photocatalyst is really difficult.[180] On this account, photoelectrochemical water splitting metrics is applied in addition to evaluate the photocatalytic activity of the Au@Nb@H$_x$K$_{1-x}$NbO$_3$ nanopeapods to reinforce the above conclusion.

In this metrics, simulated AM 1.5 G solar light coupled with diverse spectral filters was likewise employed as irradiation source. In contrast to the dye metrics, the measurements were carried out either in the two- or the three-electrode cell configuration, wherein the Au@Nb@H$_x$K$_{1-x}$NbO$_3$ nanopeapods and the H$_x$K$_{1-x}$NbO$_3$ nanoscrolls were employed as the working electrode subjected to the irradiation. The photoelectrodes were fabricated upon drop-casting 100 µL of suspension (20 g L^{-1}) containing either the Au@Nb@H$_x$K$_{1-x}$NbO$_3$ nanopeapods or the H$_x$K$_{1-x}$NbO$_3$ nanoscrolls in hexane onto a tin-doped indium oxide (ITO) conducting substrate without post-thermal and -chemical treatment. Either a platinum (Pt) coil or foil served as the counter electrode in these schemes while the three-electrode system further employed an additional silver/silver chloride reference electrode (Ag/AgCl in 3 M KCl, 0.207 V vs. NHE). An argon-purged aqueous solution containing 0.5 M sodium sulfate (Na$_2$SO$_4$) was utilized as the electrolyte. A potentiostat (CH Instruments, CHI 627D) was employed to register the current flowing from the working electrodes to the Pt counter electrode when irradiative perturbation was imposed on either the Au@Nb@H$_x$K$_{1-x}$NbO$_3$ nanopeapods or the H$_x$K$_{1-x}$NbO$_3$ nanoscrolls. Virtually, such galvanic response underlies this metrics, provided that the current amplitude substantially reflect the reaction rates of the faradaic electrolysis and the photoelectrolysis of water, respectively.[208]

The *in operando* analysis of the headspace of a custom built, air-tight, single-compartment cell in two-electrode configuration using gas chromatography (GC, Shimadzu Corporation) substantiates this argumentation (Fig. 5.22). This scheme employs topical Au@Nb@H$_x$K$_{1-x}$NbO$_3$ nanopeapods as the working electrode, which was exposed to continuous illumination of AM 1.5 G simulated sunlight for a 5-h period and anodically polarized to +1 V with regard to a Pt coil that served as the counter cathode. 100 µL of gas were sampled from the headspace of the cell every one hour using a gas-tight syringe, which was then injected into the gas-sampling loop of a GC. This GC was equipped with a packed MolSieve 5 Å column and a thermal conductivity detector (TCD). Argon (Airgas, ultra-high purity) was used as the carrier gas. The photoproduct was qualified as H$_2$ and O$_2$ gas, and the quantitative analysis suggests a stoichiometry of two H$_2$ per one O$_2$ molecule, corroborating water cleavage giving rise to gas evolution. Particularly, the quantification was made on the premise that i) the side effect of water electrolysis due to the external bias was subtracted from the measurement, and ii) the dissolution of the evolved gases in the electrolyte was calibrated using Henry's law.[209-211] To this end, additional *in operando* analysis of the equivalent system except the absence of sunlight irradiation was employed to account for the side reaction of water electrolysis.

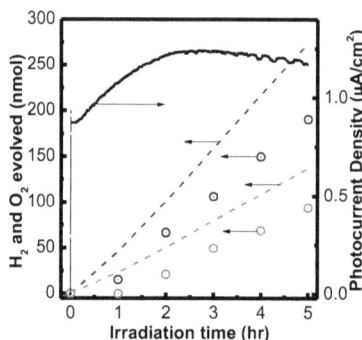

Figure 5. 22. Photocurrent transient (baseline subtracted) of the Au@Nb@H$_x$K$_{1-x}$NbO$_3$ photoelectrode polarized to +1 V (vs. Pt) under AM 1.5 G simulated sunlight illumination and the concurrent generation of H$_2$ and O$_2$. Blue and red dashed lines correspond to the integration of the net photocurrent converted into the amount of H$_2$ and O$_2$ gas using Faraday's law, respectively.[216] Blue and red circled dots correspond to experimentally measured H$_2$ and O$_2$ gas using GC, respectively.

The Faradaic efficiencies (η_F) of the H$_2$ and O$_2$ generation on the Pt cathode and the Au@Nb@H$_x$K$_{1-x}$NbO$_3$ photoanode are derived from, [212]

$$\eta_F = \frac{zFn_{molar}}{Q_{ph}}$$

(5.9)

wherein F is the Faraday constant (96485 C mol^{-1}), n_{molar} the mole amount of produced H_2 or O_2, z the number of transferred electrons per mole of evolved gas (viz. 2 e$^-$ for H_2 and 4 e$^-$ for O_2), and Q_{ph} is the integrated photogenerated charge. Eqn. 5.9 suggests η_F of ca. 0.7, reinforcing that photocurrent results primarily from faradaic water photoelectrolysis along with minor reactions that are most likely of relevance to the formation of surface peroxo-species.[212-215]

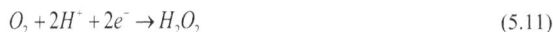

$$2H_2O \rightarrow H_2O_2 + 2H^+ + 2e^- \tag{5.10}$$

$$O_2 + 2H^+ + 2e^- \rightarrow H_2O_2 \tag{5.11}$$

On such basis, the current fluctuation was first collected for the Au@Nb@H$_x$K$_{1-x}$NbO$_3$ photoelectrode as a function of potential that was otherwise measured against the Ag/AgCl reference electrode in the three-electrode system (Fig. 5.23a).[216] A potential window from -0.25 to +1.3 V (vs. Ag/AgCl) was adopted in this measurement with a voltage sweeping rate of 10 mV s^{-1}. Chopped full solar irradiation was casted on the photoelectrode from the back side through a quartz window and the colorless electrolyte at a frequency of nearly 0.15 Hz and a fluence of 100 mW cm^{-2}, respectively.

At first glance, the Au@Nb@H$_x$K$_{1-x}$NbO$_3$ photoelectrode begins to exhibit an anodic photocurrent at ca. 0.1 V (inset in Fig. 5.23a), which then readily grows to 0.9 µA cm^{-2} at 1 V (vs. Ag/AgCl). The small photocurrent density under low external bias is ascribed to poor carrier rectification in the H$_x$K$_{1-x}$NbO$_3$ nanoscrolls due to an ineffective potential barrier established at the Schottky interface with the bimetallic Au@Nb nanocrystals and at the liquid junction with the electrolyte, respectively (Fig. 5.23b).[44] In consequence, significant charge losses via electron tunneling at the interfaces, bulk and surface recombination markedly short-circuit the photoelectrochemical cell, leading to only a modicum of the carriers being available to water photoelectrolysis.[44] Upon anodic polarization, the energy barrier particularly at the electrode/electrolyte boundary is reinforced by the applied voltage, which in turn effectively quenches undesirable shunt losses and boosts the carrier injection efficiency into the water medium (Fig. 5.23c).[213]

On this account, the temporal current fluctuations of the Au@Nb@H$_x$K$_{1-x}$NbO$_3$ and the H$_x$K$_{1-x}$NbO$_3$ photoelectrodes in response to the imposition of light are registered in the presence of a strong external bias up to +1 V (vs. Ag/AgCl). Specifically, the Au@Nb@H$_x$K$_{1-x}$NbO$_3$ photoelectrode exhibits a photocurrent that is twice the performance of the H$_x$K$_{1-x}$NbO$_3$

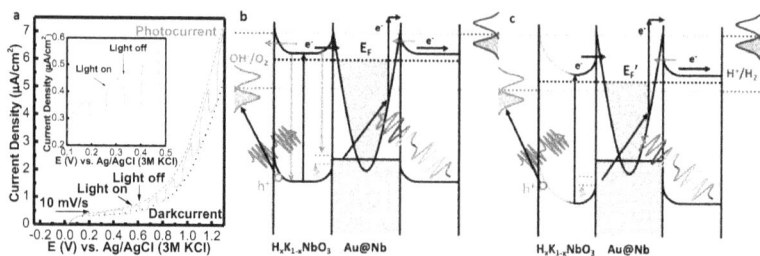

Figure 5. 23. (a) Current-potential characteristics of the Au@Nb@H$_x$K$_{1-x}$NbO$_3$ photoelectrode in 0.5 M Na$_2$SO$_4$ solution (pH 6.8) under chopped (orange line) and continuous (green dashed line) AM 1.5 G illumination. The dark current is plotted alongside (dark dotted line). (b) Characteristic shunt losses (brown arrows) including electron tunneling (brown horizontal arrows), interface and bulk charge recombination (brown vertical arrows) deteriorate the carrier collection for photoelectrochemical water splitting in the absence of an anodic bias. (c) In the presence of a strong anodic potential, the reinforced band bending within the space-charge layer (light blue curve) at the electrode/electrolyte interface effectively rectifies the charge transport. (b,c) Purple and colorful oscillations represent the UV and VIS-NIR light, respectively. Parabolic blue and underneath rectangular green hatches in the middle represent the *sp*- and *d*-bands of the plasmonic Au@Nb bimetallic, respectively. Abbreviations used: E_F and E_F', Fermi level.

photoelectrode under full sunlight illumination (Fig. 5.24), agreeing very well with the evaluation extracted from the photocatalytic dye degradation metrics (Fig. 5.19a,d,g). More importantly, the photocurrent enhancement derived from the normalization of the photocurrent of the Au@Nb@H$_x$K$_{1-x}$NbO$_3$ photoelectrode to that of the H$_x$K$_{1-x}$NbO$_3$ benchmark is highly wavelength-dependent, which progressively multiplies under illumination with longer wavelengths (Fig. 5.24). Particularly, the enhancement factor reaches three and one order of magnitude under irradiation of integral VIS-NIR and red-NIR light, respectively. Most surprisingly, the maximum improvement is derived under radiation of NIR light alone, although the photon energy (<1.8 eV) is far below those of VIS (2-3 eV) and UV (>3 eV) photons.

In addition to the reinforcement, one further needs to notice that the contour of the transient photocurrent of the Au@Nb@H$_x$K$_{1-x}$NbO$_3$ photoelectrode (left side in Fig. 5.25) is unlike that of the H$_x$K$_{1-x}$NbO$_3$ benchmark that exhibits an approximately rectangular profile in the light-on period (right side in Fig. 5.25). In contrast, the Au@Nb@H$_x$K$_{1-x}$NbO$_3$ photoelectrode demonstrates an initial current shoot (I_{in}) at the light-on instant and a steady current raise follows up along with the elapsed photoirradiation time, giving rise to the blade-like shape (left side in Fig. 5.25). Such deviation suggests a non-equilibrium between the carrier generation rate under light casting and the

98

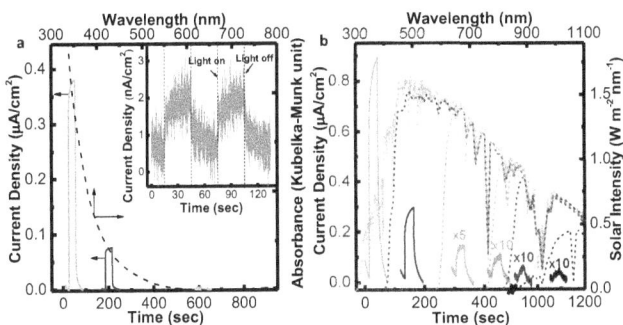

Figure 5. 24. Photocurrent-time plots (baseline subtracted) of the $H_xK_{1-x}NbO_3$ photoelectrode (a) and the $Au@Nb@H_xK_{1-x}NbO_3$ photoelectrode (b) collected in 0.5 M Na_2SO_4 solution (pH 6.8) in the presence of an anodic bias of 1 V (vs. Ag/AgCl) under irradiation of AM 1.5 G simulated sunlight (grey line) equipped with either a UV-cutoff filter (blue line) or diverse long-pass filters with cut-on wavelengths of 610 (green line), 780 (red line), 850 (brown line) and 1000 nm (black line), respectively. Kubelka-Munk transformed diffuse reflection spectrum of the $H_xK_{1-x}NbO_3$ nanoscrolls (black dashed line in a) and the spectra of AM 1.5 G simulated sunlight irradiation (grey dot line in b) after passing through either a UV-cutoff filter (blue dot line in b) or diverse long-pass filters with respective cut-on wavelengths of 610 (green dot line in b), 780 (red dot line in b), 850 (brown dot line in b) and 1000 nm (black dot line in b) are plotted alongside for reference. Inset in (a): Photocurrent-time plot (baseline subtracted) of the $H_xK_{1-x}NbO_3$ photoelectrode under chopped integral red-NIR illumination in the presence of an anodic bias of 1 V (vs. Ag/AgCl).

consumption rate upon reacting with water molecules.[213] This discrepancy may result from one or more of the following effects: i) the sluggish kinetics of water cleavage, wherein the overall reaction involves a four-electron transfer to generate H_2 and O_2 ($2H_2O \rightarrow 2H_2 + O_2$),[209] and ii) the cascade charge delivery starts with the hot electrons photogenerated on the plasmonic Au@Nb bimetallics, which then traverse the Schottky junction and migrate to $H_xK_{1-x}NbO_3$, eventually drifting to the semiconductor-liquid junction and injecting into the electrolyte.[46]

Given the challenge of water splitting lying primarily in the formidable complexity of the oxidative half-reaction ($2H_2O \rightarrow O_2 + 4H^+ + 4e^-$), 1 M sodium formate (Na(COOH)) is added to accelerate the reaction kinetics. The oxidation of formate is kinetically favorable in view of the one-electron process ([HCOO]$^- \rightarrow CO_2 + H^+ + e^-$),[217] which thus accounts for the nearly doubled photocurrent density of the $Au@Nb@H_xK_{1-x}NbO_3$ photoelectrode (Fig. 5.26) in comparison with that in the absence of Na(COOH) (Fig. 5.24b). Nevertheless, the transient photocurrent is yet in the blade-like form, which decisively rules out the responsibility of poor water oxidation kinetics for the disequilibrium of the carrier formation and depletion rates.

99

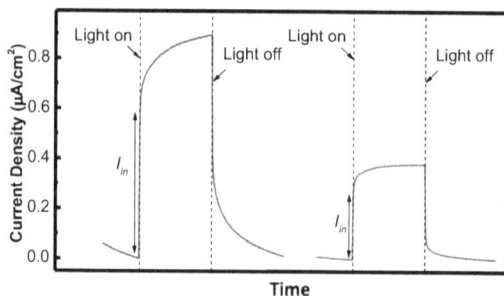

Figure 5. 25. Photocurrent-time plots (baseline subtracted) of the $H_xK_{1-x}NbO_3$ (right side) and the $Au@Nb@H_xK_{1-x}NbO_3$ photoelectrodes (left side) collected in 0.5 M Na_2SO_4 solution (pH 6.8) in the presence of an anodic bias of 1 V (vs. Ag/AgCl).

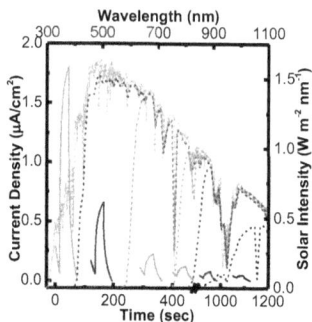

Figure 5. 26. Temporal photocurrent responses of the $Au@Nb@H_xK_{1-x}NbO_3$ photoelectrode collected in the presence of an anodic bias of 1 V (vs. Ag/AgCl) in 0.5 M Na_2SO_4 solution along with 1 M Na(COOH) hole scavenger under irradiation of AM 1.5 G simulated sunlight (grey line) equipped with either a UV-cutoff filter (blue line) or diverse long-pass filters with cut-on wavelengths of 610 (green line), 780 (red line), 850 (brown line) and 1000 nm (black line), respectively. The spectra of AM 1.5 G simulated sunlight irradiation (grey dot line) after passing through either a UV-cutoff filter (blue dot line) or diverse long-pass filters with respective cut-on wavelengths of 610 (green dot line), 780 (red dot line), 850 (brown dot line) and 1000 nm (black dot line) are plotted alongside for reference.

With broadband sunlight illumination, both the $H_xK_{1-x}NbO_3$ semiconductor and the plasmonic Au@Nb bimetallic are photoexcited and charge carriers are formed (Fig. 5.27). Particularly, energetic electrons (e_{sp}^-) of the plasmonic Au@Nb bimetallic immediately populate $H_xK_{1-x}NbO_3$ via the DET mechanism.[202,203] The anodic bias then dictates those majority charges (e_{sp}^- and e_{CB}^-) moving to the Pt counter electrode for water reduction to H_2.[218] In contrast, the minority holes (h_{VB}^+) of the $H_xK_{1-x}NbO_3$ semiconductor are scavenged by the surrounding aqueous medium to generate O_2 at the Au@Nb@$H_xK_{1-x}NbO_3$ photoelectrode. The overall process is completed when a second UV photon further excites the $H_xK_{1-x}NbO_3$ semiconductor to produce the next e_{CB}^-/h_{VB}^+ pair and an electron transfer follows up to compensate the accumulated holes (h_{sp}^+) on the core-shell Au@Nb nanocrystals implanted in the $H_xK_{1-x}NbO_3$ nanoscrolls.[219,220]

The elementary steps making up this kinetic scheme are formulated below

$$hv + Au@Nb \xrightarrow{\;k_1 F_{hv}\;} e_{sp}^- + h_{sp}^+ \tag{5.12}$$

$$hv + H_xK_{1-x}NbO_3 \xrightarrow{\;k_2 F_{hv}\;} e_{CB}^- + h_{VB}^+ \tag{5.13}$$

$$e_{CB}^- + h_{sp}^+ \xrightarrow{\;k_3\;} heat \tag{5.14}$$

$$h_{VB}^+ + H_2O \xrightarrow{\;k_4\;} \frac{1}{2}O_2 + 2H^+ \tag{5.15}$$

Figure 5. 27. Energy band diagram and the charge transfer underlying water photooxidation by the Au@Nb@$H_xK_{1-x}NbO_3$ photoelectrode in the presence of an anodic bias of 1 V (vs. Ag/AgCl) under irradiation of AM 1.5 G simulated sunlight at a fluence of 100 mW cm^{-2}. Purple and colorful oscillations represent the UV and VIS-NIR light, respectively. Parabolic blue and underneath rectangular green hatches in the middle represent the *sp*- and *d*-bands of the plasmonic Au@Nb bimetallic, respectively. Abbreviations used: e_{CB}^-, electron in CB of the $H_xK_{1-x}NbO_3$ semiconductor; h_{VB}^+, hole in VB of the $H_xK_{1-x}NbO_3$ semiconductor; e_{sp}^-, electron in the *sp*-band of the plasmonic Au@Nb bimetallic; h_{sp}^+, hole in the *sp*-band of the plasmonic Au@Nb bimetallic; E_F, Fermi level.

Herein, F_{hv} is the photon flux impinging on the Au@Nb@H$_x$K$_{1-x}$NbO$_3$ photoelectrode, h_{sp}^+ and e_{sp}^- the hole and the electron in the sp-band of the core-shell Au@Nb bimetallic, h_{VB}^+ and e_{CB}^- are the hole and electron in VB and CB of the H$_x$K$_{1-x}$NbO$_3$ semiconductor, respectively.[213,221,222] This model suggests a linear dependence of the hole flux on the light intensity for the Au@Nb@H$_x$K$_{1-x}$NbO$_3$ photoelectrode in the absence of charge-carrier losses.

$$-\frac{d[h_{VB}^+]}{dt} = -\frac{d[H_2O]}{dt} = k_2 k_4 [H_2O] F_{hv} \tag{5.16}$$

Such dependency turns into a square-root-type relation

$$-\frac{d[h_{VB}^+]}{dt} = -\frac{d[H_2O]}{dt} = k_4 [H_2O] (\frac{k_2 k_3 - k_1 k_6}{k_3 k_5})^{\frac{1}{2}} F_{hv}^{\frac{1}{2}} \tag{5.17}$$

in the presence of bulk recombination of geminate charge carriers.[208,221]

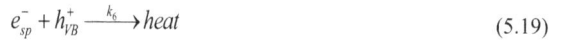

$$e_{CB}^- + h_{VB}^+ \xrightarrow{k_5} heat \tag{5.18}$$

$$e_{sp}^- + h_{VB}^+ \xrightarrow{k_6} heat \tag{5.19}$$

Eqn. 5.17 is derived on the premise of i) a fast equilibrium between the charge-carrier formation (Eqn. 5.12 and Eqn. 5.13) and the recombination (Eqn. 5.14, Eqn. 5.18 and Eqn. 5.19), and ii) the steady-state approximation of $[e_{sp}^-] \approx [h_{sp}^+]$ = constant and $[e_{CB}^-] \approx [h_{VB}^+]$ = constant. Specifically, the hole flux reaching the surface of the Au@Nb@H$_x$K$_{1-x}$NbO$_3$ photoelectrode equals the rate of water photooxidation (Eqn. 5.15 and Eqn. 5.17) that is faithfully manifested in the experimentally measured anodic photocurrent density (Fig. 5.24b). Noteworthily, such derivation is true only i) in the absence of surface recombination, and ii) at low F_{hv} characterized by $[h_{VB}^+] \ll [H_2O]$.[213] More importantly, the transient photocurrent of the Au@Nb@H$_x$K$_{1-x}$NbO$_3$ photoelectrode measured in the presence of an anodic bias of 1V (vs. Ag/AgCl) under irradiation of AM 1.5 G simulated sunlight at a fluence of 100 mW cm^{-2} excellently meets such demands. A very important consequence of this strong anodic bias is the significantly reinforced potential barrier (V_B) within the space-charge layer, leading to a virtually "frozen out" charge recombination at the electrode/electrolyte interface owing to the majority electrons that nearly deplet in the space-charge region, as suggested by the Boltzmann expression.[33,34,48,213,223]

$$n_s = n_b \exp\left(-\frac{qV_B}{k_B T}\right) \tag{5.20}$$

In Eqn. 5.20, n_s and n_b represent the surface and bulk density of electrons, q the electronic charge, k_B the Boltzmann constant, and T is the temperature. On such basis, the collected photocurrent density substantially describes the hole flux that participates in the faradaic water photooxidation. Moreover, the photocurrent transients are measured at the exposure to the broadband sunlight irradiation with the intensity not more than "one sun" (viz. 100 mW cm^{-2}) to ensure the reaction rate mostly dictated by the hole flux in preference to the water oxidation kinetics.[213]

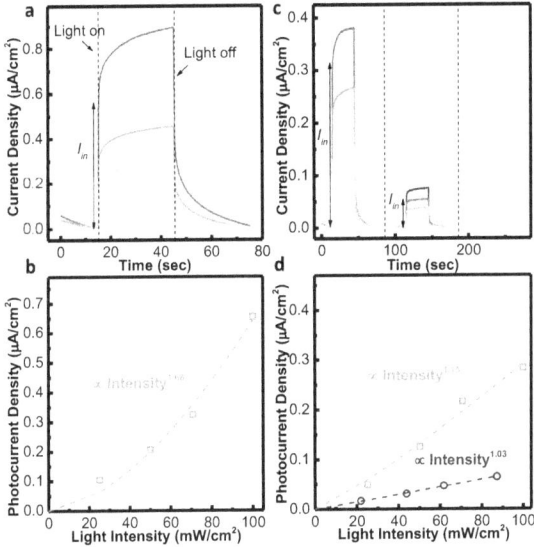

Figure 5. 28. Fluence-dependent photocurrent-time plots (baseline subtracted) of the Au@Nb@H$_x$K$_{1-x}$NbO$_3$ (a) and the H$_x$K$_{1-x}$NbO$_3$ (b) photoelectrodes, respectively. The initial photocurrent shoot (I_{in}) as a function of the intensity of either simulated sunlight (grey line) or integral VIS-NIR light (blue line) for the Au@Nb@H$_x$K$_{1-x}$NbO$_3$ (c) and H$_x$K$_{1-x}$NbO$_3$ (d) photoelectrodes, respectively. The intensity of these broadband irradiations is gradually attenuated (top to bottom) using various graduated neutral-density filters.

Surprisingly, in the fluence-dependent temporal photocurrent measurements (Fig. 5.28) the Au@Nb@H$_x$K$_{1-x}$NbO$_3$ photoelectrode demonstrates a superlinear dependency of I_{in} (Fig. 5.28a) on the intensity of AM 1.5 G simulated sunlight (Fig. 5.28b). This suggests the presence of a nonlinear optical effect on the hole flux of the Au@Nb@H$_x$K$_{1-x}$NbO$_3$ photoelectrode. Moreover, the quasi-linear dependency (Fig. 5.28d) of I_{in} of the H$_x$K$_{1-x}$NbO$_3$ photoelectrode (Fig. 5.28c) on the intensity of AM 1.5 G simulated sunlight highlights that such nonlinear optics exclusively influences the hole flow photogenerated by the bimetallic Au@Nb nanocrystals via the LSPR excitation. Moreover, such discrepancy in fluence-dependency between the Au@Nb@H$_x$K$_{1-x}$NbO$_3$ and H$_x$K$_{1-x}$NbO$_3$ photoelectrodes (Fig. 5.28b and Fig. 5.28d) punctuates that those plasmonic charges are the mainstream of the overall hole current. Most importantly, the pseudo-quadratic correlation implies a two-photon-assisted charge generation mechanism of the centrosymmetric Au@Nb bimetallic via a third-order photon-photon interaction. This is in good agreement with characteristic nonlinearity of the well-documented two sequential one-photon absorption process in the literature.[222,224]

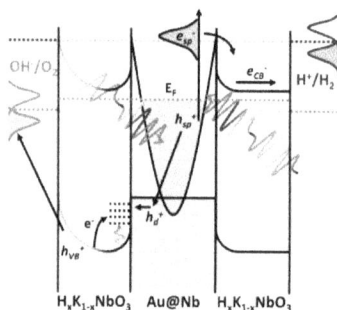

Figure 5. 29. Energy band diagram and the charge transfer building on nonlinear plasmonics, which underlies water photooxidation by the Au@Nb@H$_x$K$_{1-x}$NbO$_3$ photoelectrode in the presence of an anodic bias of 1 V (vs. Ag/AgCl) under irradiation of AM 1.5 G simulated sunlight at a fluence of 100 mW cm^{-2}. Colorful and flamy oscillations represent the VIS-NIR light and the phonon, respectively. Parabolic blue and underneath rectangular green hatches in the middle represent the sp- and d-bands of the plasmonic Au@Nb bimetallic, respectively. Dash lines in the bandgap of the H$_x$K$_{1-x}$NbO$_3$ semiconductor represent the defect-associated surface states (SS). Abbreviations used: e_{CB}^-, electron in CB of the H$_x$K$_{1-x}$NbO$_3$ semiconductor; h_{VB}^+, hole in VB of the H$_x$K$_{1-x}$NbO$_3$ semiconductor; e_{sp}^-, electron in the sp-band of the plasmonic Au@Nb bimetallic; h_{sp}^+, hole in the sp-band of the plasmonic Au@Nb bimetallic; h_d^+, hole in the d-band of the plasmonic Au@Nb bimetallic; E_F, Fermi level.

The elementary steps building on such nonlinear plasmonics are described below (Fig. 5.29).

$$hv + Au@Nb \xrightarrow{\;k_1F_{hv}\;} e_{sp}^- + h_{sp}^+ \qquad (5.12)$$

$$hv + h_{sp}^+ \xrightarrow{k_7 F_{hv}} h_d^+ \tag{5.21}$$

$$h_d^+ + SS \xrightarrow{k_8} SS^+ \tag{5.22}$$

$$SS^+ + \hbar\omega \xrightarrow{k_9 F_{\hbar\omega}} h_{VB}^+ \tag{5.23}$$

$$h_{VB}^+ + H_2O \xrightarrow{k_4} \frac{1}{2}O_2 + 2H^+ \tag{5.15}$$

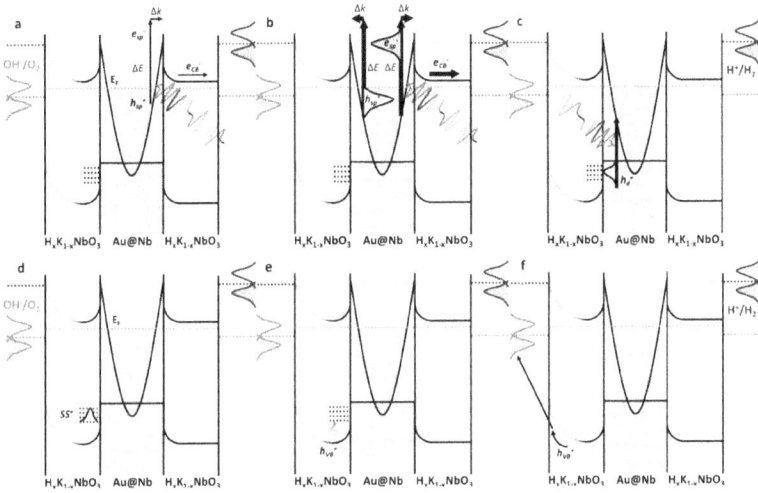

Figure 5. 30. Elementary steps making up the two sequential one-photon absorption process (a-c) and the charge transfer (a-f) building on such nonlinear plasmonics, which underlies the water photooxidation by the Au@Nb@H$_x$K$_{1-x}$NbO$_3$ photoelectrode polarized to 1 V (vs. Ag/AgCl) under irradiation of AM 1.5 G simulated sunlight at a fluence of 100 mW cm^{-2}. Colorful and flamy oscillations represent the VIS-NIR light and the phonon, respectively. Parabolic blue and underneath rectangular green hatches represent the *sp*- and *d*-bands of the plasmonic Au@Nb bimetallic, respectively. Dash lines in the bandgap of the H$_x$K$_{1-x}$NbO$_3$ semiconductor represent the defect-associated surface states (*SS*). Abbreviations used: Δk, momentum change; ΔE, energy change; e_{CB}^-, electron in CB of the H$_x$K$_{1-x}$NbO$_3$ semiconductor; h_{VB}^+, hole in VB of the H$_x$K$_{1-x}$NbO$_3$ semiconductor; e_{sp}^-, electron in the *sp*-band of the plasmonic Au@Nb bimetallic; h_{sp}^+, hole in the *sp*-band of the plasmonic Au@Nb bimetallic; h_d^+, hole in the *d*-band of the plasmonic Au@Nb bimetallic; E_F, Fermi level.

Herein, h_d^+ is the hole in the d-band of the core-shell Au@Nb bimetallic, $F_{h\omega}$ the phonon flux spontaneously emitted from the lattice vibration, and SS^+ is the hole at the mid-gap surface state of the $H_xK_{1-x}NbO_3$ nanoscrolls at the Schottky interface.[222]

In this formulism, the process always begins with the formation of a first hot electron (e_{sp}^-) of plasmonic Au@Nb bimetallic via an intraband $sp \rightarrow sp$ transition through the LSPR excitation (Eqn. 5.12). Such electronic transition is indirect in nature and requires changes not only in energy (ΔE) but also in momentum (Δk). Particularly, ΔE is exclusively harvested from the decay of the surface plasmon polaritons (SPPs) that are generated by resonant photons (insets in Fig. 5.18). In contrast, Δk is more likely derived with the aid of a phonon or a lattice imperfection in addition to the wavevector (k-vector) of the incident photon (Fig. 5.30a and Fig. 5.31a,b).[204,225,226]

Such defect-assisted hot electron-hole pair generation mechanism suggests that energetic charges (e_{sp}^- and h_{sp}^+) are preferentially formed at the margin of a sub-10 nm Au@Nb nanocrystal in view of the numerous, intrinsic undercoordination sites available at the boundary with the $H_xK_{1-x}NbO_3$ nanoscrolls (Fig. 5.31b). Such selective spatial distribution, and moreover, the three-dimensional Schottky interface (Fig. 5.31a,b) that originates from the specific nanopeapod configuration (Fig. 4.13) greatly favor hot electron injection from plasmonic Au@Nb bimetallics to the $H_xK_{1-x}NbO_3$ semiconductor.[227,228] Moreover, the abundant available DOS in CB of the $H_xK_{1-x}NbO_3$ semiconductor, which primarily builds on unoccupied Nb $4d$ states (Fig. 5.31e), endows $H_xK_{1-x}NbO_3$ with excellent electron-uptake properties. This further supports the electron injection.

On account of the vacant sp-band of the Au core of the Au@Nb bimetallics building on the hybridization of empty Au $5d/6s$-p orbitals and the d^0 configuration ([Kr]$4d^0 5s^0$) of unoccupied CB of the $H_xK_{1-x}NbO_3$ nanoscrolls, in-operando XAS at the Au and Nb L_3-edge is employed in the present contribution to study this DET journey (Fig. 5.31c,d). Herein, "in-operando" literally refers to measurements performed on the Au@Nb@$H_xK_{1-x}NbO_3$ nanopeapods in the presence of simulated AM 1.5 G sunlight irradiation. The collected spectrum is then in comparison with that in the absence of solar illumination. At first sight, an evident absorbance increment is manifested in the in-operando Au L_3-edge XAFS of the Au@Nb@$H_xK_{1-x}NbO_3$ nanopeapods with respect to that of normal Au L_3-edge XAFS collected without solar irradiation (Fig. 5.31c). This suggests that additional electron vacancies are available to the electronic transition from the Au $2p$ to hybridized s-p-d states via the X-ray excitation (left panel in Fig. 5.31e). In contrast to the Au L_3-edge XAS analysis, an absorbance decrement is manifested in the in-operando Nb L_3-edge XAFS of the Au@Nb@$H_xK_{1-x}NbO_3$ nanopeapods with respect to that of normal Nb L_3-edge XAFS collected

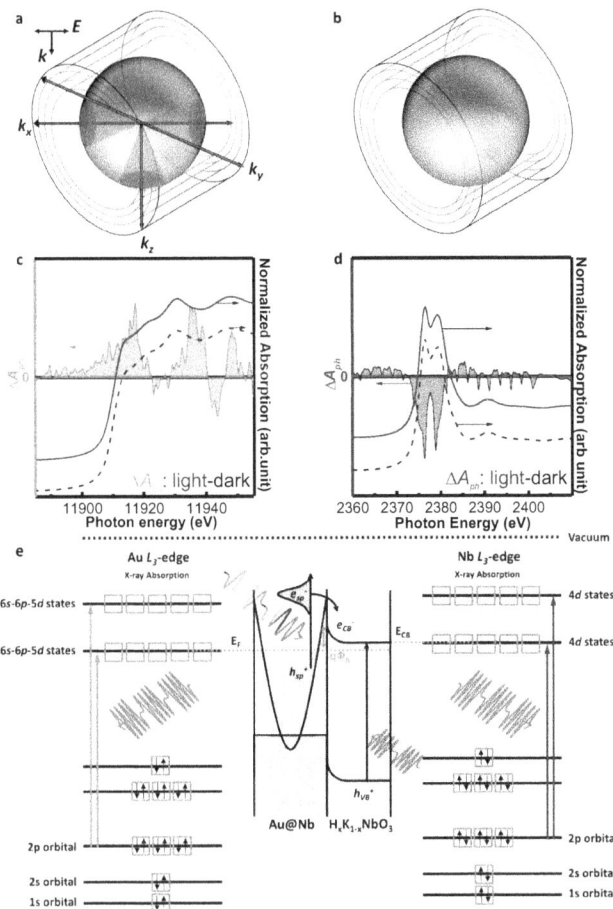

Figure 5. 31. (a,b) Schematic illustrations of the characteristic three-dimensional Schottky junction of the Au@Nb@H$_x$K$_{1-x}$NbO$_3$ nanopeapods. Hot electrons derive the momentum from both (a) the k-vector of the incident photon and (b) the surface imperfection sites (the red margin of the colorful sphere in (b)) of plasmonic Au@Nb bimetallic. Electrons then traverse the Schottky junction only if (a) their k-vector lies inside the emission cones and (e) their energy exceeds the Schottky barrier height ($q\Phi_B$). Normalized (c) Au and (d) Nb L_3-edge XAF structures of the Au@Nb@H$_x$K$_{1-x}$NbO$_3$ nanopeapods collected in the presence (blue line) and absence (black dash line) of AM 1.5 G simulated sunlight. The difference in absorbance (ΔA_{ph}) substantially reflects (e) the charge transfer of the hot electrons of the core-shell Au@Nb bimetallic into CB of the H$_x$K$_{1-x}$NbO$_3$ semiconductor. Grey, colorful and purple oscillations in (e) represent the X-ray, UV and VIS-NIR light, respectively. Parabolic blue and underneath rectangular green hatches in the middle represent the

sp- and d-bands of the plasmonic Au@Nb bimetallic, respectively. Abbreviations used: e_{CB}^-, electron in CB of the $H_xK_{1-x}NbO_3$ semiconductor; h_{VB}^+, hole in VB of the $H_xK_{1-x}NbO_3$ semiconductor; e_{sp}^-, electron in the sp-band of the plasmonic Au@Nb bimetallic; h_{sp}^+, hole in the sp-band of the plasmonic Au@Nb bimetallic; E_F, Fermi level; $q\Phi_B$, Schottky barrier height.

exclusive of sunlight illumination (Fig. 5.31d). This also implies that the electronic transition from the Nb $2p$ to the vacant $4d$ states is suppressed (right panel in Fig. 5.31e). Altogether, such evidential differences in absorbance variation between the Au and the Nb L_3-edge XAFS effectively validates the utility of DET to transform the e_{sp}^- into e_{CB}^- (Fig. 5.30a,b).[229] For plasmonic d^0 metal oxide composites, DET virtually takes place on a femtoseconds timeframe.[230] Moreover, the electron transfer is devoid of energy dissipation in terms of the characteristic mean free path (MFP) up to the order of few tens of nanometres, which is ascribed to the free-electron-like nature of the sp-band.[231]

Subsequent to this process is another LSPR excitation via the next incoming photon that produces additional hot carriers, including either i) an energetic e_{sp}^-/h_{sp}^+ pair via indirect intraband $sp \rightarrow sp$ transition (Eqn. 5.12), or ii) lukewarm e_{sp}^- coupled with hot h_d^+ via direct interband $d \rightarrow sp$ transition (Eqn. 5.12 and Eqn. 5.21). In the first few femtoseconds, the former process predominates in view of the ample DOS above the Fermi level,[225] leading to a rapid accumulation of tepid h_{sp}^+ with energies below the Fermi level (Fig. 5.30b).[232] Herein, the external bias plays an important role of immediately shuttling e_{CB}^- to the Pt counter electrode, which effectively quenches the annihilation of h_{sp}^+ by e_{CB}^- via the charge recombination.[33,34] This in turn results in a quantitative surge of h_{sp}^+, consequently reinforcing the probability of the interband $d \rightarrow sp$ transition, wherein lukewarm h_{sp}^+ are substantially restored to generate hot h_d^+ (Fig. 5.30c). Such two sequential one-photon absorption (Fig. 5.30a-c) should not be overlooked in the present study, provided that the Au@Nb@$H_xK_{1-x}NbO_3$ nanopeapods demonstrate a strong broadband absorption from visible up to NIR light via the LSPR excitations (Fig. 5.18). Moreover, the consequent localized surface plasmons (LSPs) significantly boost the local electromagnetic field (insets in Fig. 5.18), further promoting this nonlinear plasmonics.[222,224]

The absence of momentum change (Δk) in the direct interband $d \rightarrow sp$ transition suggests the delocalization of hot h_d^+.[225,232] Unlike the free-electron-like properties of the sp-band, the d-band is bound in nature owing to the high DOS otherwise suggesting significant electron-phonon scattering.[225,230,233] In consequence, the featured MFP of the d-band carriers is of the order of very few nanometres.[225,231,234] In other words, only h_d^+ formed within the first few nanometres at the surface of plasmonic Au@Nb bimetallics highly likely traverse the boundary into the $H_xK_{1-x}NbO_3$

nanoscrolls at the absence of energy dissipation (Fig. 5.30d). Such issue is more-or-less mitigated in the present study in view of the enormous surface-area-to-volume ratio of the sub-10 nm Au@Nb nanocrystals, suggesting presumably more than three tenths of h_d^+ entering $H_xK_{1-x}NbO_3$, which builds on the premise of MFP ≥ 1 nm for h_d^+ at the d-band edge.[231] Particularly, the mid-gap surface states (SS) of the $H_xK_{1-x}NbO_3$ nanoscrolls at the interface with the core-shell Au@Nb bimetallics function as important relays in this isoenergetic hole uptake (Eqn. 5.22).[34,203,213]

Most SS stem highly likely from the stepwise nano-texturization in the retrosynthesis of the $H_xK_{1-x}NbO_3$ nanoscrolls (Fig. 4.10). Specifically, those defect-associated mid-gap electronic levels are in general identified with a shallow energy position close to the VB maximum.[63] This in turn suggests the feasibility of a phonon-assisted transformation of the holes at those trapped sites (SS^+) into h_{VB}^+ (Fig. 5.30e). Such premise is corroborated by the evidential photocurrent transient of the $H_xK_{1-x}NbO_3$ photoelectrode under irradiation of sub-bandgap integral VIS-NIR light (Fig. 5.24a). This temporal photocurrent is presumably attributed to a sequential process starting with an electronic transition either from discrete surface-defect level to the continuous band state or *vice versa*. Afterwards, h^+ or e^- at the mid-gap energy state ends up in entering the VB or CB via a thermal transfer. Eventually, the strong upward band bending (V_B) in the depletion layer of the $H_xK_{1-x}NbO_3$ nanoscrolls (Fig. 5.30f), which results from the anodic bias and which is oriented toward the water phase, drives h_{VB}^+ to the solid-liquid boundary and injects into the electrolyte for O_2 generation (Eqn. 5.15).

The overall kinetic analysis suggests a quadratic dependence

$$-\frac{d[h_{VB}^+]}{dt} = -\frac{d[H_2O]}{dt} = k_1 k_4 k_7 k_8 k_9 [SS][H_2O]F_{h\omega}F_{hv}^2 \qquad (5.24)$$

of the hole flux of the Au@Nb@$H_xK_{1-x}NbO_3$ photoelectrode on the light intensity in the absence of bulk charge-carrier losses and scattering. The disagreement with the measured sub-quadratic correlation (Fig. 5.28b) suggests that the charge-carrier decay should not be overlooked in the present contribution. The bulk recombination of e_{CB}^- and h_{VB}^+ of the $H_xK_{1-x}NbO_3$ nanoscrolls is virtually ruled out, provided that the featured quasi-linear dependence (Fig. 5.28d) of I_{in} of the $H_xK_{1-x}NbO_3$ photoelectrode (Fig. 5.28c) on the intensity of AM 1.5 G simulated sunlight implies that such losses are effectively quenched. This is attributed to the depletion region developed via the applied bias nearly overwhelming the overall $H_xK_{1-x}NbO_3$ nanoscrolls, provided that the characteristic width (w) is dictated by the extent of V_B (Eqn. 2.8).[33,34] In other words, most charge carriers are lost in the process of the LSPR excitations, wherein the charge scattering and recombination are the major contributors.[222,225,231] Particularly, the scattering of h_{sp}^+ and h_d^+ take place at a rate of $[h_{sp}^+]/\tau_{sp}$ and $[h_d^+]/\tau_d$, wherein τ_{sp} and τ_d represent the relaxation time of the sp- and d-band holes.

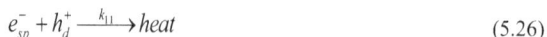

$$e_{sp}^- + h_{sp}^+ \xrightarrow{k_{10}} heat \tag{5.25}$$

$$e_{sp}^- + h_d^+ \xrightarrow{k_{11}} heat \tag{5.26}$$

τ is in general on a femtosecond timeframe, suggesting that h_{sp}^+ and h_d^+ very likely undergo scattering events prior to the recombination. The inclusion of such carrier losses (Eqn. 5.25 and Eqn. 5.26) in the kinetic interpretation gives rise to a dependency of the photooxidation rate of H_2O on the sub-second-order incident light intensity.

$$-\frac{d[H_2O]}{dt} = k_4 k_8 k_9 [SS][H_2O]F_{h\nu} \frac{\dfrac{-k_7}{2k_{10}\tau_{sp}}F_{h\nu} + \dfrac{-k_7}{2k_{10}}F_{h\nu}\sqrt{\dfrac{1}{\tau_{sp}^2} + 4k_1 k_{10} F_{h\nu}}}{\left(k_8[SS] + \dfrac{1}{\tau_d} - \dfrac{k_{11}}{2k_{10}\tau_{sp}}\right) + \dfrac{k_{11}}{2k_{10}}\sqrt{\dfrac{1}{\tau_{sp}^2} + 4k_1 k_{10} F_{h\nu}}} \tag{5.27}$$

Particularly, Eqn. 5.27 is derived on the premise that the probability of the recombination between e_{sp}^- and h_{sp}^+ (Eqn. 5.25) predominates over that between e_{sp}^- and h_d^+ (Eqn. 5.26) in view of $[h_{sp}^+]$ $\gg$ $[h_d^+]$ (Fig. 5.30b,c). On such basis, $[e_{sp}^-] = [h_{sp}^+] + [h_d^+] \approx [h_{sp}^+]$ is likewise concluded. Additionally, the assumption of i) the fast equilibrium between the charge-carrier formation (Eqn. 5.12 and Eqn. 5.21), scattering and recombination (Eqn. 5.25 and Eqn. 5.26), and ii) the steady-state approximation of $[e_{sp}^-] \approx [h_{sp}^+]$ = constant are also applied. This derivation satisfactorily accounts for the sub-quadratic dependency (Fig. 5.28b) of I_{in} of the Au@Nb@H$_x$K$_{1-x}$NbO$_3$ photoelectrode (Fig. 5.28a) on the intensity of AM 1.5 G simulated sunlight.

In addition to preceding mathematical insight, the photoelectric conversion efficiency summarized in the action spectrum (Fig. 5.32) reveals further details of the characteristic nonlinear plasmonics of the Au@Nb@H$_x$K$_{1-x}$NbO$_3$ photoelectrode. The incident photon to charge carrier efficiency (IPCE) is derived via dividing the number of photogenerated electrons of the Au@Nb@H$_x$K$_{1-x}$NbO$_3$ photoelectrode by the number of the incident photons (Eqn. 5.28).[235]

$$IPCE = \frac{I_{ph}}{P_{in}} \times \frac{1240}{\lambda} \tag{5.28}$$

In this expression, I_{ph} is the photocurrent density (A cm^{-2}) of the Au@Nb@H$_x$K$_{1-x}$NbO$_3$ photoelectrode, P_{in} and λ are the power density (W cm^{-2}) and characteristic wavelength (nm) of the incoming photon, respectively. Discrete monochromatic light with the wavelengths spanned from ca. 400 up to nearly 1000 nm, a width of $\pm$ 5 nm and the intensity ranged from 20 to 50 mW cm^{-2} are employed in this measurement and the Au@Nb@H$_x$K$_{1-x}$NbO$_3$ photoelectrode is anodically polarized to +1 V (vs. Ag/AgCl). Prima facie, the photocurrent action spectrum of the Au@Nb@H$_x$K$_{1-x}$NbO$_3$ photoelectrode tracks the transverse-mode LSPR band and longitudinal-mode LSPR plateau that manifest in the Kubelka-Munk transformed diffuse reflectance spectra (Fig. 5.15b) of the Au@Nb@H$_x$K$_{1-x}$NbO$_3$ nanopeapods reasonably well. Such agreement corroborates that the VIS- and NIR-light-active photoelectrochemical water is substantially triggered by the plasmonic charge of the implanted bimetallic Au@Nb nanoparticles. The IPCE gradually lowers from the blue to the red end of the action spectrum (Fig. 5.32), suggesting that the sequential intraband $sp \rightarrow sp$ and interband $d \rightarrow sp$ transitions via the two sequential one-photon absorption process are mostly triggered by the visible photons with fractional aids of the red-NIR photons.

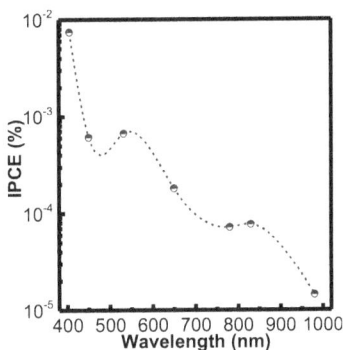

Figure 5. 32. Photocurrent action spectrum of the Au@Nb@H$_x$K$_{1-x}$NbO$_3$ photoelectrode polarized to +1 V (vs. Ag/AgCl) under discrete monochromatic irradiations.

Virtually, IPCE collected in this VIS-NIR wavelength region is substantially of order of a few to few tens of millionth. Under full solar illumination at a fluence of 100 mW cm^{-2}, this in turn gives rise to the photocurrent density at the level of few tenths to a few nanoamperes (nA cm^{-2}) at the individual monochromatic wavelengths. Given the VIS and NIR light exhibiting a wavelength span of hundreds to thousands of nanometers in the diffusive sunlight,[236] the integrated photocurrent density over this interval is up to the order of few tenths of microamperes (μA cm^{-2}).

This is in excellent agreement in the order of magnitude with that of the photocurrent transients of the Au@Nb@$H_xK_{1-x}NbO_3$ photoelectrode derived under sunlight illumination exclusive of the UV light (Fig. 5.24b). Such trace level of IPCE value also indicates appreciable shunt losses via the charge scattering and recombination (Eqn. 5.25 and Eqn. 5.26) in the present nonlinear process, which in turn significantly thermalizes the ionic lattice of the plasmonic Au@Nb bimetallics.[196,225]

Hence, the presence of refractory Nb shells and the $H_xK_{1-x}NbO_3$ nanosheaths effectively quench the dissolution and agglomeration of the Au@Nb nanocrystals. Statistical analysis over hundreds of TEM images suggests a minor fusion of closely neighbored Au@Nb nanoparticles (inset in Fig. 5.33a) upon the development of the Nb neck in between (Fig. 5.33b), giving rise to the formation of bimetallic Au@Nb nanopeanuts characterized by a slightly larger diameter of 21 ± 7 nm (Fig. 5.33a).

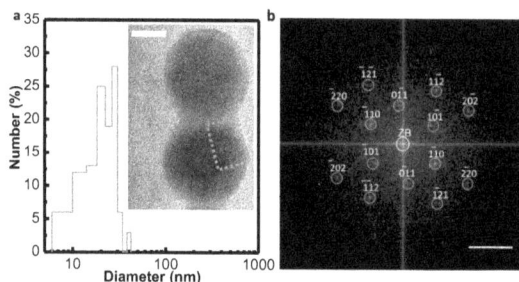

Figure 5. 33. (a) Statistical analysis of the particle diameter of bimetallic Au@Nb nanocrystals in peanut-like form. Representative TEM image (scale bar: 7 nm) is shown in the inset. (b) Diffractogram (scale bar: 7 nm^{-1}) of the core-shell Au@Nb nanopeanut on the neck region (blue frame in the inset in (a)) and calculated diffraction pattern with Miller indices of bulk cubic Nb (space group *Im-3m*, a = 3.32 Å) in the [115]-zone axis. The white circle in (b) indicates the zero-order beam (ZB).

More importantly, the constant presence of bimetallic Au@Nb nanoparticles (inset in Fig. 5.33 and inset in Fig. 5.34) suggests the feasibility of long-term photoelectrochemical water splitting (Fig. 5.34) because the excitation of LSPRs by the visible and NIR photons are yet highly efficient. In this long-term solar water cleavage by the Au@Nb@$H_xK_{1-x}NbO_3$ photoelectrode, the minor deterioration of the photocurrent and the gas evolving rate over elapsed photoelectrolysis period can be most likely ascribed to the progressive attenuated IPCE value along with the red-shift of the resonant wavelength of LSPRs. Such resonant transition results presumably from the morphological evolution of plasmonic Au@Nb bimetallic from the centrosymmetric nanospeheres (Fig. 5.17a) to the nanopeanuts characterized by higher aspect ratio (inset in Fig. 5.33a).[196]

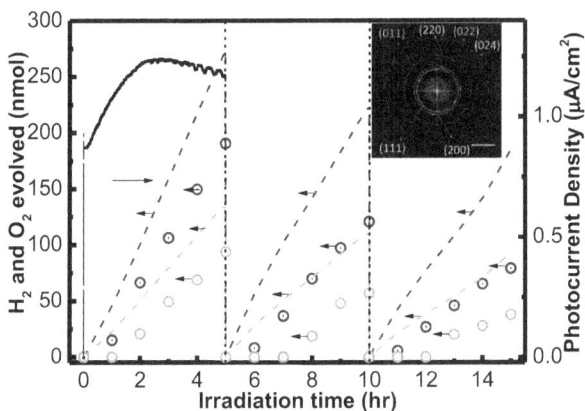

Figure 5. 34. Long-term photocurrent transient (baseline subtracted) of the Au@Nb@H$_x$K$_{1-x}$NbO$_3$ photoelectrode polarized at +1 V (vs. Pt) under AM 1.5 G simulated sunlight illumination and the concurrent generation of H$_2$ and O$_2$. Blue and red dashed lines correspond to the integration of the net photocurrent converted into the amount of H$_2$ and O$_2$ gas using Faraday's law, respectively.[216] Blue and red circled dots correspond to experimentally measured H$_2$ and O$_2$ gas using GC, respectively. Inset: SAED pattern of the core-shell Au@Nb nanopeanuts formed after the long-term water photoelectrolysis, wherein the characteristic Deby-Sherrer rings of (orange circles) bulk *fcc* Au (space group *Fm-3m*, a = 4.08 Å) and (blue circles) bulk cubic Nb (space group *Im-3m*, a = 3.32 Å) are yet concurrently present.

In summary, the biomimicry of the peapod-like nanoarchitecture endows Au@Nb@H$_x$K$_{1-x}$NbO$_3$ with an exceptional light harvesting ability well befitting the solar spectrum. Particularly, the outmost H$_x$K$_{1-x}$NbO$_3$ nanosheaths and the inmost Au@Nb nanoparticles are responsible for the UV light and most VIS light absorption, respectively. More importantly, the nanoscale resolution of the bimetallic Au@Nb chain that unidirectionally extends to a sub-microscopic length stimulates the near-field plasmon-plasmon coupling between adjacent bimetallic entities. Such dipolar interaction accounts for the photoabsorption of Au@Nb@H$_x$K$_{1-x}$NbO$_3$ toward the NIR light. Last but not least, the dye photodegradation and water photoelectrolysis metrics validates the superior VIS- and NIR-light-active photocatalytic properties of the Au@Nb@H$_x$K$_{1-x}$NbO$_3$ nanopeapods. Inclusive chemical transformations are exclusively triggered by the charge-carriers of the bimetallic Au@Nb nanoparticles photogenerated via the nonlinear plasmonics.

6. Summary

The aim of this dissertation is directed toward subduing the frontier of material design via putting forward two modern nanoarchitectures – spikecubes and nanopeapods – for sunlight-driven photoelectrochemical applications. Particularly, the spikecubes exemplified by β-SnWO$_4$ are named for the first time after the peculiar morphology, wherein the self-aligned, quasi-periodic nanopillar array builds on six sharp faces of a hexahedral microcubic core. In contrast, the nanopeapods typified by Au@Nb@H$_x$K$_{1-x}$NbO$_3$ are called after the biomimetic peapod-like blueprint, wherein semi-infinite, discrete sub-10 nm Au@Nb nanoparticles (peas) are uniaxially implanted in the cavity of tubular H$_x$K$_{1-x}$NbO$_3$ nanoscrolls (pods).

Virtually, both the β-SnWO$_4$ spikecubes and the Au@Nb@H$_x$K$_{1-x}$NbO$_3$ nanopeapods are prepared in the context of the bottom-up approach. A polyol synthesis is employed for the β-SnWO$_4$ spikecubes formation and a soft-chemical multistep process is adopted to prepare the Au@Nb@H$_x$K$_{1-x}$NbO$_3$ nanopeapods. The crystal growth of the β-SnWO$_4$ spikecubes starts with the formation of spherical nanoparticles with narrow size distribution, which then thermally transform into hexagonal microcubes enclosed by six sharp {100} facets via the thermodynamically-driven Ostwald ripening process. Subsequently, the growth scheme is hydrodynamically transited to a kinetic mode, wherein the preformed microcrystals serve as the seeds for the columnar deposition of β-SnWO$_4$ to form surface nanospikes. The β-SnWO$_4$ spikecubes is characterized by a particle size of 2-9 μm for the underlying microcube, an arm length of 0.7-2 μm, and a base diameter of 200 nm for the anisotropic nanospike, respectively.

In addition, the soft-chemical multi-phase synthesis of the Au@Nb@H$_x$K$_{1-x}$NbO$_3$ nanopeapods starts with a soft-chemical solid-state retrosynthesis of a cubic H$_x$K$_{1-x}$NbO$_3$ perovskite from parental H$_x$K$_{4-x}$Nb$_6$O$_{17}$ characterized by a layered crystal structure. Spontaneous nanoorigami takes place along with the retrosynthesis, leading to the final H$_x$K$_{1-x}$NbO$_3$ nanoscrolls adopting the anisotropic tubule form of the H$_x$K$_{4-x}$Nb$_6$O$_{17}$ precursor. Subsequently, a common solution reduction technique is employed to integrate the Au nanocrystals into the preformed H$_x$K$_{1-x}$NbO$_3$ nanoscrolls. Particularly, the spatially inhomogeneous distribution of OAm reducing agent results in most Au nanocrystals implanted discretely inside the cavity of the H$_x$K$_{1-x}$NbO$_3$ nanoscrolls. Such configuration promotes the eventual deposition of Nb refractory metal on the surface of preformed Au nanocrystals at underpotentials. The Au@Nb@H$_x$K$_{1-x}$NbO$_3$ nanopeapods are characterized by a particle size of 9 nm for the inmost core-shell Au@Nb bimetallic, a thickness of 1 nm for the Nb shell, an interparticle distance of 2 nm for the discrete bimetallic Au@Nb chain, and a sub- to microscopic length of the $H_xK_{1-x}NbO_3$ nanoscrolls.

The significance of crystallizing β-$SnWO_4$ into spikecubes and preparing Au@Nb@$H_xK_{1-x}NbO_3$ in the biomimetic form of nanopeapods for the photoelectrocatalytic functionality are next evaluated by the dye photodegradation and the water photoelectrolysis metrics. The characteristic open framework and multibranched structure of the spikecubes effectively reinforces the surface reaction sites of β-$SnWO_4$. Moreover, the undercoordinated atoms at these active sites significantly reframe the band energetics of β-$SnWO_4$ via the distortion-mediated SOJT effect. In consequence, the CB minimum of the β-$SnWO_4$ spikecubes is at a negative energy level (on scale of V vs. NHE), further boosting the redox power toward the RhB and MB dye molecules. Altogether, the synergistic effect of this morphological and textural engineering accounts for the superior photocatalytic activity of the β-$SnWO_4$ spikecubes, which well outperforms that of fine-particulate WO_3 and microcubic β-$SnWO_4$ benchmarks by more than 150% and 243%, respectively.

The integration of plasmonic Au@Nb nanoparticles into semiconducting $H_xK_{1-x}NbO_3$ nanoscrolls renders the outstanding photocatalytic properties of the Au@Nb@$H_xK_{1-x}NbO_3$ nanopeapods not only UV- but also VIS-light active. Moreover, the biomimicry of the peapod-like nanoarchitecture stimulates the near-field plasmon-plasmon coupling between the adjacent bimetallic Au@Nb entities. Such dipolar interaction further endows Au@Nb@$H_xK_{1-x}NbO_3$ with an exceptional NIR-light-active photocatalytic characteristic. Particularly, the chemical transformation involved in the VIS- and NIR-light-triggered dye photodegradation and water photoelectrolysis are exclusively mediated by the charge-carriers of bimetallic Au@Nb nanoparticles photogenerated via the nonlinear plasmonics. Last but not least, the characteristic three-dimensional Schottky junction of the Au@Nb@$H_xK_{1-x}NbO_3$ nanopeapods greatly favors the delivery of these plasmonic charges.

Altogether, a promising "one-pot" polyol-mediated synthesis has been successfully developed, whereby several design criteria for photocatalysts including high surface-to-volume ratio, open framework with facile accessibility, and optimal band structure, have been simultaneously addressed. In particular, the thermal effect in such synthetic context, which grabbed only scarce attention to date, has turned out for the first time playing the key role in this comprehensive fulfillment. In addition, in this dissertation a biomimetic blueprint has been put forward for the first time to cater for an additional design requisite for photocatalysts, viz. the light harvesting ability. Particularly, the realization of full-spectrum utilization of diffusive sunlight, including not only the well-exploited UV and VIS light but also the most abundant NIR light that has currently been often overlooked, renders this design exceptionally appealing for a large variety of photoactive devices.

7. Outlook

The inspiration of chemical utilizations of renewable solar energy by natural plants has stimulated a tremendous surge of research activities in the field of photoelectrochemistry worldwide. This field, which makes use of the principles of photochemistry, electrochemistry, and semiconductor physical chemistry, has nurtured the development and design of next-generation photoelectrochemical cells and photocatalysts. Although numerous past researches have accumulated significant advances, which originate from single-crystal semiconductor photoelectrodes in the 1960s and which have expanded to nanostructured counterparts and the photodiodes nowadays, neither scheme meets the end of practical solar engineering. Thus, from the point of view of industrialization and commercialization, many challenges remain in the areas of materials science and engineering.

While the benefits of the nanoscaling approach have been well-documented over the past decade, which has been in favor of the charge transfer due to increased electroactive interfaces with the redox medium, has improved the light distribution via strong scattering, and promoted the redox power via the quantum confinement effect, challenges are simultaneously imposed on the charge collection and transportation. Thus, minimizing the shunt losses that are mostly present at the grain and phase boundaries are highly desirable. To address these issues, efforts to tailor the crystal growth need to be further pursued in order to prepare nanostructured materials with high quality. Additionally, tandem configurations are likewise a well promising alternative in view of the excellent rectifying properties due to the built-in electric field at the heterojunction. Moreover, such design allows the composites to more efficiently take advantage of diffusive sunlight. Albeit in this category plasmonic device lag behind the Z-scheme systems in terms of the overall power conversion efficiency, many salient features offer opportunities for the improvement. Hence, endeavors to manipulate the metal-metal interaction and the metal-semiconductor interface need to be continually proceeded.

To meet a substantial breakthrough, systematic efforts in not only material preparation and device fabrication but also in characterizing the surface and interface properties at the atomic level are indispensable. Particularly, in-situ analyses are highly informative, provided that such techniques offer a real picture of the interfacial charge transfer in action. Moreover, computational material screening and photocatalytic process simulation are likewise valuable, given that these studies alternatively provide theoretical clues to searching for novel highly efficient photoelectrode and photocatalyst materials. The next decade will continue to see such interdisciplinary synergy stimulating a burgeoning growth in the photoelectrochemical field.

8. Literature

[1]   Benyus, J. M. *Biomimicry: Innovation Inspired by Nature*, 1st Ed., Harper Perennial, New York (1997).

[2]   Becquerel, E. *C. R. Acad. Sci.* 1839, **9**, 561.

[3]   Brattain, W. H., Garrett, C. G. B. *Bell Syst. Tech. J.* 1955, **34**, 129.

[4]   Delahay, P. *Advances in Electrochemistry and Electrochemical Engineering*, 1st Ed., Wiley-Interscience, New York (1961)

[5]   Eyring, H., Henderson, D., Jost, W. *Physical Chemistry: An Advanced Treatise*, Academic Press, New York (1970).

[6]   Miamlin, V. A., Pleskov, Y. V. *Electrochemistry of Semiconductors,* 1st Ed., Springer-Verlag, New York (1967).

[7]   Vijh, A. K. *Electrochemistry of Metals and Semiconductors: The Application of Solid State Science to Electrochemical Phenomena*, Marcel Dekker, New York (1973).

[8]   Hannay, N. B. *Semiconductors*, Reinhold Pub. Co., New York (1959).

[9]   Green, M. In *Modern Aspects of Electrochemistry*, Ed. Bockris, J. O'M., Butterworths, London (1959).

[10]   Morrison, S. R. *Prog. Surf. Sci.* 1971, **1**, 105.

[11]   Efimov, E. A., Erusalimchik, I. G. *Electrochemistry of Semiconductors*, Sigma Press, Washington (1963).

[12]   Gerischer, H. *Z. Phys. Chem.* 1960, **26**, 325.

[13]   Gerischer, H. *Z. Phys. Chem.* 1961, **27**, 48.

[14]   Gerischer, H. *J. Electrochem. Soc.* 1966, **113**, 1174.

[15]   Gerischer, H. *Surf. Sci.* 1969, **18**, 97.

[16]   Turner, D. R. *J. Electrochem. Soc.* 19556, **103**, 252.

[17]   Turner, D. R. *J. Electrochem. Soc.* 1958, **105**, 40.

[18]   Boddy, P. J. *J. Electrochem. Soc.* 1968, **115**, 199.

[19]   Boddy, P. J., Kahng, D., Chen, Y. S. *Electrochim. Acta* 1968, **13**, 1311.

[20]   Dewald, J. F. *J. Phys. Chem. Solids* 1961, **14**, 155.

[21]   Williams, R. *J. Chem. Phys.* 1960, **32**, 1505.

[22]   Morrison, S. R. *Surf. Sci.* 1969, **15**, 363.

[23]   Morrison, S. R., Freund, T. *J. Chem. Phys.* 1967, **47**, 1543.

[24]   Gomes, W. P., Freund, T., Morrison, S. R. *Surf. Sci.* 1969, **13**, 201.

[25]   Gomes, W. P., Freund, T., Morrison, S. R. *J. Electrochem. Soc.* 1968, **115**, 818.

[26]   Freund, T., Morrison, S. R. *Surf. Sci.* 1968, **9**, 119.

[27]   Memming, R. *J. Electrochem. Soc.* 1969, **116**, 785.

[28]   Memming, R., Schwandt, G. *Electrochim. Acta* 1968, **13**, 1299.

[29]   Fujishima, A., Honda, K. *Nature* 1972, **238**, 37.

Literature

[30]   Halmann, M. *Nature* 1978, **275**, 115.
[31]   Dickson, C. R., Nozik, A. J. *J. Am. Chem. Soc.* 1978, **100**, 8007.
[32]   Frank, S. N., Bard, A. J. *J. Am. Chem. Soc.* 1977, **99**, 4667.
[33]   Bard, A. J. *J. Photochem.* 1979, **10**, 59.
[34]   Nozik, A. J. *Annu. Rev. Phys. Chem.* 1978, **29**, 189.
[35]   Duonghong, D., Borgarello, E., Grätzel, M. *J. Am. Chem. Soc.* 1981, **103**, 4685.
[36]   Dimitrijevic, N. M., Li, S., Grätzel, M. *J. Am. Chem. Soc.* 1984, **106**, 6565.
[37]   Hagfeldt, A., Grätzel, M. *Chem. Rev.* 1995, **95**, 49.
[38]   Rossetti, R., Ellison, J. L., Gibson, J. M., Brus, L. E. *J. Chem. Phys.* 1984, **80**, 4464.
[39]   Fojtik, A., Weller, H., Koch, U., Henglein, A. *Ber. Bunseiges. Phys. Chem.* 1984, **88**, 969.
[40]   Yoffe, A. D. *Adv. Phys.* 2001, **50**, 1.
[41]   Kudo, A., Miseki, Y. *Chem. Soc. Rev.* 2009, **38**, 253.
[42]   Kamat, P. V., Tvrdy, K., Baker, D. R., Radich, J. G. *Chem. Rev.* 2010, **110**, 6664.
[43]   Tong, H., Ouyang, S., Bi, Y., Umezawa, N., Oshikiri, M., Ye, J. *Adv. Mater.* 2012, **24**, 229.
[44]   Osterloh, F. E. *Chem. Soc. Rev.* 2013, **42**, 2294.
[45]   Chen, Y. C., Lin, Y. G., Hsu, L. C., Tarasov, A., Chen, P. T., Hayashi, M., Ungelenk, J., Hsu, Y. K., Feldmann, C. *ACS Catal.* 2016, **6**, 2357.
[46]   Chen, Y. C., Hsu, Y. K., Popescu, R., Gerthsen, D., Lin, Y. G., Feldmann, C. *Nature Commun.* 2018, **9**, 232.
[47]   Xu, Y., Schoonen, M. A. A. *Am. Mineral.* 2000, **85**, 543.
[48]   Gelderman, K., Lee, L., Donne, S. W. *J. Chem. Educ.* 2007, **84**, 685.
[49]   Bockris, J.O'M., Khan, S. U. M. *Surface electrochemistry: a molecular level approach*, Plenum Press, New York (1993).
[50]   Butler, M. A., Ginley, D. S. *J. Electrochem. Soc.* 1978, **125**, 228.
[51]   Halouani, F. E., Deschavres, A. *Mater. Res. Bull.* 1982, **17**, 1045.
[52]   Matsumoto, Y., Yoshikawa, T., Sato, E. *J. Electrochem. Soc.* 1989, **136**, 1389.
[53]   Morrison, S. R. *The Chemical Physics of Surfaces*, Plenum, 1st Ed., New York (1977).
[54]   Seraphin, B. O. *Solar Energy Conversion: Solid-State Physics Aspects*, 1st Ed., Springer-Verlag, Berlin Heidelberg (1979).
[55]   L. I. Berger, in *CRC Handbook of Chemistry and Physics*, Ed. D. R. Lide, CRC Press/Taylor and Francis, Boca Raton, FL (2008).
[56]   Borg, R. J., Dienes, G. J. *The Physical Chemistry of Solids*, Academic Press, Boston (1992).
[57]   Shuey, R.T. *Semiconducting Ore Minerals*, 1st Ed., Elsevier Scientific Pub. Co., Amsterdam (1975).
[58]   Vaughan, D. J., Craig, J. R. *Mineral Chemistry of Metal Sulphides*, Cambridge University Press, Cambridge (1978).

[59]   Vogel, R., Hoyer, P., Weller, H. *J. Phys. Chem.* 1994, **98**, 3183.

[60]   Kronik, L., Ashkenasy, N., Leibovitch, M., Fefer, E., Yoram, S., Gorer, S., Hodes, G. *J. Electrochem. Soc.* 1998, **145**, 1748.

[61]   Tvrdy, K., Frantsuzov, P. A., Kamat, P. V. *Proc. Natl. Acad. Sci. U. S. A.* 2011, **108**, 29.

[62]   Holmes, M. A., Townsend, T. K., Osterloh, F. E. *Chem. Commun.* 2012, **48**, 371.

[63]   Kubacka, A., Fernández-García, M., Colón, G. *Chem. Rev.* 2012, **112**, 1555.

[64]   Coronado, J. M., Sánchez, B., Portela, R., Suárez, S. *J. Sol. Energy Eng.* 2008, **130**, 011016.

[65]   Kandiel, T. A., Feldhoff, A., Robben, L., Dillert, R., Bahnemann, D. W., *Chem. Mater.* 2010, **22**, 2050.

[66]   Paola, A. D., Bellardita, M., Ceccato, R., Palmisano, L., Parrino, F. *J. Phys. Chem. C* 2009, **113**, 15166.

[67]   Kavan, L., Grätzel, M., Gilbert, S. E., Klemenz, C., Scheel, H. J. *J. Am. Chem. Soc.* 1996, **118**, 6716.

[68]   Leland, J. K., Bard, A. J. *J. Phys. Chem.* 1987, **91**, 5076.

[69]   Lin, Y. C., Dumcenco, D. O., Huang, Y. S., Suenaga, K. *Nature Nanotechnol.* 2014, **9**, 391.

[70]   Zhao, M., Xu, H., Chen, H., Ouyang, S., Umezawa, N., Wang, D., Ye, J. *J. Mater. Chem. A* 2015, **3**, 2331.

[71]   Karpinski, A., Berson, S., Terrisse, H., Granvalet, M. M., Guillerez, S., Brohan, L., Richard-Plouet, M. *Sol. Energy Mater Sol. Cells* 2013, **116**, 27.

[72]   Schaak, R. E., Mallouk, T. E. *Chem. Mater.* 2002, **14**, 1455.

[73]   Nicolosi, V., Chhowalla, M., Kanatzidis, M. G., Strano, M. S., Coleman, J. N., *Science* 2013, **340**, 1226419.

[74]   Schaak, R. E., Mallouk, T. E. *Chem. Mater.* 2000, **12**, 3427.

[75]   Sabio, E. M., Chamousis, R. L., Browning, N. D., Osterloh, F. E. *J. Phys. Chem. C* 2012, **116**, 3161.

[76]   Ebina, Y., Sasaki, T., Harada, M., Watanabe, M. *Chem. Mater.* 2002, **14**, 4390.

[77]   Murphy, A. B. *Int. J. Hydrogen Energy* 2006, 31, 1999.

[78]   Youngblood, W. J., Lee, S. A., Maeda, K., Mallouk, T. E. *Acc. Chem. Res.* 2009, **42**, 1966.

[79]   Linic, S., Christopher, P., Ingram, D. B. *Nature Mater.* 2011, **10**, 911.

[80]   Kamat, P. V. *J. Phys. Chem. Lett.* 2012, **3**, 663.

[81]   Walter, M. G., Warren, E. L., McKone, J. R., Boettcher, S. W., Mi, Q. X., Santori, E. A., Lewis, N. S. *Chem. Rev.* 2010, **110**, 6446.

[82]   Lewis, N. S. *Inorg. Chem.* 2005, **44**, 6900.

[83]   Crooker, S. A., Hollingsworth, J. A., Tretiak, S., Klimov, V. I. *Phys. Rev. Lett.*, 2002, **89**, 186802.

[84]   Koole, R., Liljeroth, P., Donegá, C. M., Vanmaekelbergh, D., Meijerink, A. *J. Am. Chem. Soc.* 2006, 128, 10436.

[85]   Tong, H., Umezawa, N., Ye, J. *Chem. Commun.* 2011, **47**, 4219.

[86]   Tong, H., Umezawa, N., Ye, J., Ohno, T. *Energy Environ. Sci.* 2011, **4**, 1684.

[87]   Cheng, J. Y., Ross, C. A., Chan, V. Z. H., Thomas, E. L., Lammertink, R. G. H., Vancso, G. J. *Adv. Mater.* 2001, **13**, 1174.

[88]   Ling, X. Y., Phang, I. Y., Maijenburg, W., Schonherr, H., Reinhoudt, D. N., Vancso, G. J., Huskens, J. *Angew. Chem., Int. Ed.* 2009, **48**, 983.

[89]   Weaver, J. H., Waddill, G. D. *Science* 1991, **251**, 1444.

[90]   Korth, B. D., Keng, P., Shim, I., Bowles, S. E., Tang, C., Kowalewski, T., Nebesny, K. W., Pyun, J. *J. Am. Chem. Soc.* 2006, **128**, 6562.

[91]   Tao, A. R., Huang, J., Yang, P. *Acc. Chem. Res.* 2008, **41**, 1662.

[92]   Aldaye, F. A., Palmer, A. L., Sleiman, H. F. *Science* 2008, **321**, 1795.

[93]   Srivastava, S., Santos, A., Critchley, K., Kim, K. S., Podsiadlo, P., Sun, K., Lee, J., Xu, C., Lilly, G. D., Glotzer, S. C., Kotov, N. A. *Science* 2010, **327**, 1355.

[94]   Kongkanand, A., Tvrdy, K., Takechi, K., Kuno, M., Kamat, P. V. *J. Am. Chem. Soc.* 2008, **130**, 4007.

[95]   Kao, J., Thorkelsson, K., Bai, P., Rancatore, B. J., Xu, T. *Chem. Soc. Rev.* 2013, **42**, 2654.

[96]   Hatton, B., Mishchenko, L., Davis, S., Sandhage, K. H., Aizenberg, J. *Proc. Natl. Acad. Sci. U. S. A.* 2010, **107**, 10354.

[97]   Chou, L. Y. T., Zagorovsky, K., Chan, W. C. W. *Nature Nanotechnol.*, 2014, **9**, 148.

[98]   Liu, N., Mukherjee, S., Bao, K., Brown, L. V., Dorfmüller, J., Nordlander, P., Halas, N. J. *Nano Lett.* 2012, **12**, 364.

[99]   Lin, Y. G., Hsu, Y. K., Chen, Y. C., Chen, L. C., Chen, S. Y., Chen, K. H. *Nanoscale* 2012, **4**, 6515.

[100]   Cong, V. T., Ganbold, E., Saha, J. K., Jang, J., Min, J., Choo, J., Kim, S., Song, N. W., Son, S. J., Lee, S. B., Joo, S. W. *J. Am. Chem. Soc.* 2014, **136**, 3833.

[101]   Sears, F. W. *Optics*, 3rd Ed., Addison-Wesley Pub. Co., Cambridge Massachusetts (1949).

[102]   Fultz, B., Howe, J. *Transmission Electron Microscopy and Diffractometry of Materials*, 4th Ed., Springer-Verlag, Berlin Heidelberg (2008).

[103]   Als-Nielsen, J., McMorrow, D. *Elements of Modern X-ray Physics*, 1st Ed., John Wiley & Sons, Chichester (2001).

[104]   Warren, B.E. *X-ray Diffraction*, 1st Ed., Addison-Wesley Pub. Co., Massachusetts (1969).

[105]   Cullity, B.D. *Elements of X-Ray Diffraction*, 2nd Ed., Addison-Wesley Pub. Co., Massachusetts (1978).

[106]   Waseda, Y., Matsubara, E., Shinoda, K. *X-Ray Diffraction Crystallography: Introduction, Examples and Solved Problems*, Springer-Verlag, Berlin Heidelberg (2011).

[107]   Walsh, A., Payne, D. J., Egdell, R. G., Watson, G. W. *Chem. Soc. Rev.* 2011, **40**, 4455.

[108] ICDD, *International Center for Diffraction Data*, 2003.

[109] Douglas, A. S. *Fundamentals of Analytical Chemistry*, 8th Ed., Thomson-Brooks/Cole, Belmont, CA (2004).

[110] Kaufmann, E. N. *Characterization of Materials*, 1st Ed., John Wiley & Sons, Hoboken, NJ (2003).

[111] Owen, T. *Fundamentals of Modern UV-Visible Spectroscopy: A Primer*, Hewlett-Packard Company (1996).

[112] Bohren, C. F., Huffman, D. R. *Absorption and Scattering of Light by Small Particles*, John Wiley & Sons, New York (1983).

[113] Skoog, D. A., Holler, F. J., Crouch, S. R. *Principles of Instrumental Analysis*, 6th Ed., Thomson Brooks/Cole, Belmont, CA (2007).

[114] Ulery, A. L., Drees, L. R. *Methods of Soil Analysis: Part 5—Mineralogical Methods*, Soil Science Society of America, Inc., Madison, WI (2008).

[115] Tauc, J., Grigorovici, R., Vancu, A. *Phys. Status Solidi* 1966, **15**, 627.

[116] F. Abelès, *Optical Properties of Solids*, North-Holland Pub. Co., Amsterdam (1972).

[117] Davis, E. A., Mott, N. F. *Philos. Mag.* 1970, **22**, 903.

[118] Kodre, A., Arčon, I., Gomilšek, J. P. *Acta Chim. Slov.* 2004, **51**, 1.

[119] Newville, M. *Rev. Mineral. Geochem.* 2014, **78**, 33.

[120] Young, N. A. *Coord. Chem. Rev.* 2014, 277–278, 224.

[121] Yano, J., Yachandra, V. K. *Photosynth. Res.* 2009, 102, 241.

[122] Kawaguchi, T., Fukuda, K., Matsubara, E. *J. Phys.: Condens. Matter* 2017, **29**, 113002.

[123] Koch, E. –E. *Handbook on Synchrotron Radiation*, North-Holland Pub. Co., Amsterdam (1983).

[124] Stöhr, J. *NEXAFS Spectroscopy*, 1st Ed., Springer-Verlag, Berlin Heidelberg (1992).

[125] Konningsberger, D. C., Prins, R. *X-ray absorption, principles, applications, techniques of EXAFS, SEXAFS and XANES*, 1st Ed., John Wiley & Sons, New York, (1988).

[126] Schnohr, C. S., Ridgway, M. C., *X-Ray Absorption Spectroscopy of Semiconductors*, 1st Ed., Springer-Verlag, Berlin Heidelberg (2015).

[127] Borfecchia, E., Garino, C., Salassa, L., Lamberti, C. *Phil. Trans. R. Soc. A* 2013, **371**, 20120132.

[128] Solomon, D., Lehmann, J., Kinyangi, J., Liang, B., Schäfer, T. *Soil Sci. Soc. Am. J.* 2005, **69**, 107.

[129] Tsang, K. L., Lee, C. -H., Jean, Y. C., Dann, T .-E. Chen, J. R., D'Amico, K. L., Oversluizen, T. *Rev. Sci. Instrum.* 1995, **66**, 1812.

[130] Newville, M. *J. Synchrotron Rad.* 2001, **8**, 322.

[131] Chung, S. C., Chen, C. I., Tseng, P. C., Lin, H. F., Dann, T. E., Song, Y. F., Huang, L. R., Chen, C. C., Chuang, J. M., Tsang, K. L., Chang, C. N. *Rev. Sci. Instrum.* 1995, **66**, 1655.

[132] Ankudinov, A. L., Ravel, B., Rehr, J. J., Conradson, S. D. *Phys. Rev. B* 1998, **58**, 7565.

[133]   Rehr, J. J., Albers, R. C. *Rev. Mod. Phys.* 2000, **72**, 621.

[134]   Dann, T. –E., Chung, S. –C., Huang, L. –J., Juang, J. –M., Chen, C. –I., Tsang, K. –L. *J. Synchrotron Rad.* 1998, **5**, 664.

[135]   Song, Y. F., Chang, C. H., Liu, C.Y., Huang, L. J., Chang, S. H., Chuang, J. M., Chung, S. C., Tseng, P. C., Lee, J. F., Tsang, K. L., Liang, K. S. *AIP Conf. Proc.* 2004, **705**, 412.

[136]   L. Rayleigh, *Philos. Mag.* 1881, 12, 81.

[137]   Meyers, R.A. *Encyclopedia of Analytical Chemistry*, 1st Ed., John Wiley & Sons, New York (2000).

[138]   Berne, B. J., Pecora, R. *Dynamic Light Scattering: With Applications to Chemistry, Biology, and Physics*, 1st Ed., John Wiley & Sons, New York (1976).

[139]   Malvern Instruments Ltd., *Zetasizer Nano Series User Manual*, 2004.

[140]   Malvern Instruments Ltd., *A Basic Guide to Particle Characterization*, 2012.

[141]   Goodhew, P. J., Humphreys, J., Beanland, R. *Electron Microscopy and Analysis*, 3rd Ed., Taylor & Francis, London (2000).

[142]   Zhou, W., Wang, Z. L. *Scanning Microscopy for Nanotechnology: Techniques and Applications*, 1st Ed., Springer-Verlag, New York (2007).

[143]   Dong, H., Chen, Y.-C., Feldmann, C. *Green Chem.* 2015, **17**, 4107.

[144]   Kuntz, E., Kuntz, H. *Hepatology: Textbook and Atlas*, 3rd Ed., Springer-Verlag, Berlin Heidelberg (2008).

[145]   Bock, C., Paquet, C., Couillard, M., Botton, G. A., MacDougall, B. R. *J. Am. Chem. Soc.* 2004, **126**, 8028.

[146]   Xia, Y., Xiong, Y., Lim, B., Skrabalak, S. E. *Angew. Chem. Int. Ed.* 2009, **48**, 60.

[147]   Jeitschko, W., Sleight, A. W. *Acta Crystallogr. B* 1972, **28**, 3174.

[148]   Stoltzfus, M. W., Woodward, P. M., Seshadri, R., Klepeis, J. -H., Bursten, B. *Inorg. Chem.* 2007, **46**, 3839.

[149]   Wojcik, J., Calvayrac, F., Goutenoire, F., Mhadhbi, N., Corbel, G., Lacorre P., Bulou, A. *J. Phys. Chem. C* 2013, **117**, 5301.

[150]   Gilmer, G. H., Huang, H., Roland, C. *Comput. Mater. Sci.*, 1998, **12**, 354.

[151]   Chen, J., Herricks, T., Xia, Y., *Angew. Chem. Int. Ed.* 2005, **44**, 2589.

[152]   Cheetham, A. K., Day, P. *Solid State Chemistry: Techniques*, Clarendon Press, Oxford (1987).

[153]   Rao, C. N. R., Gopalakrishnan, J. *New Directions in Solid State Chemistry*, 2nd Ed., Cambridge University Press, Cambridge (1997).

[154]   Feng, S., Xu, R. *Acc. Chem. Res.* 2001, **34**, 239.

[155]   Livage, J., Henry, M., Sanchez, C. *Prog. Solid State Chem.* 1988, **18**, 259.

[156]   Jolivet, J.-P. *Metal Oxide Chemistry and Synthesis: From Solution to Solid State*, John Wiley, New York (2000).

[157]   Saupe, G. B., Waraksa, C. C., Kim, H. N., Han, Y. J., Kaschak, D. M., Skinner, D. M., Mallouk, T. E. *Chem. Mater.* 2000, **12**, 1556.

[158]   Tournoux, M., Marchand, R., Brohan, L. *Prog. Inorg. Solid State Chem.* 1986, **17**, 33.

[159]   Figlarz, M. *Chem. Scr.* 1988, **28**, 3.

[160]   Figlarz, M., Gérand, B., Delahaye-Vidal, A., Dumont, B., Harb, F., Coucou, A., Fievet, F. *Solid State Ionics* 1990, **43**, 143.

[161]   Figlarz, M., Gerand, B., Dumont, B., Delahayevidal, A., Portemer, F. *Phase Transitions* 1991, **31**, 167.

[162]   Stein, A., Keller, S. W., Mallouk, T. E. *Science* 1993, **259**, 1558.

[163]   Schleich, D. M. *Solid State Ionics* 1994, **70**, 407.

[164]   Gopalakrishnan, J. *Chem. Mater.* 1995, **7**, 1265.

[165]   Rouxel, J., Tournoux, M. *Solid State Ionics* 1996, **84**, 141.

[166]   Pribošič, I., Makovec, D., Drofenik, M. *Chem. Mater.* 2005, **17**, 2953.

[167]   Adireddy, S., Yao, Y., He, J., Wiley, J. B. *Mater. Res. Bull.* 2013, **48**, 3236.

[168]   Nassau, K., Shiever, J. W., Bernstein, J. L. *J. Electrochem. Soc.* 1969, **116**, 348.

[169]   Kong, X., Hu, D., Wen, P., Ishii, T., Tanakab, Y., Feng, Q. *Dalton Trans.*, 2013, **42**, 7699.

[170]   Fourquet, J. L., Renou, M. F., Pape, R. D., Theveneau, H., Man, P. P., Lucas, O., Pannetier, J. *Solid State Ionics* 1983, **9&10**, 1011.

[171]   Kolb, D. M., Przasnyski, M., Gerischer, H. *J. Electroanal. Chem. Interfacial Electrochem.* 1974, **54**, 25.

[172]   Sudha, V., Sangaranarayanan, M. V. *J. Phys. Chem. B* 2002, **106**, 2699.

[173]   Sudha, V., Sangaranarayanan, M. V. *J. Phys. Chem. B* 2003, **107**, 3907.

[174]   Gilroy, K. D., Ruditskiy, A., Peng, H. C., Qin, D., Xia, Y. *Chem. Rev.* 2016, **116**, 10414.

[175]   Yu, Y., Zhang, Q., Xie, J., Lee, J. Y. *Nature Commun.* 2013, **4**, 1454.

[176]   Adireddy, S., Carbo, C. E., Rostamzadeh, T., Vargas, J. M., Spinu, L., Wiley, J. B. *Angew. Chem. Int. Ed.* 2014, **53**, 4614.

[177]   Borges, J., Ribeiro, J. A., Pereira, E. M., Carreira, C. A., Pereira, C. M., Silva, F. *J. Colloid Interface Sci.* 2011, **358**, 626.

[178]   Barnard, A. S., Young, N. P., Kirkland, A. I., van Huis, M. A., Xu, H. *ACS Nano* 2009, **3**, 1431.

[179]   Wang, D., Li, Y. *J. Am. Chem. Soc.* 2010, **132**, 6280.

[180]   Ohtani, B. *Chem. Lett.* 2008, **37**, 217.

[181]   Ungelenk, J., Feldmann, C. *Appl. Catal. B* 2012, **127**, 11.

[182]   Herrmann, J. M. *Appl. Catal. B* 2010, **99**, 461.

[183]   Abe, R., Takami, H., Murakami, N., Ohtani, B. *J. Am. Chem. Soc.* 2008, **130**, 7780.

[184]   Demchenko, I. N., Denlinger, J. D., Chernyshova, M., Yu, K. M., Speaks, D. T., Olalde-Velasco, P., Hemmers, O., Walukiewicz, W., Derkachova, A., Lawniczak-Jablonska, K. *Phys. Rev. B* 2010, **82**, 07510.

[185]   Tang, J., Ye, J. *J. Mater. Chem.* 2005, **15**, 4246.

[186]   Ping, Y., Li, Y., Gygi, F., Galli, G. *Chem. Mater.* 2012, **24**, 4252.

[187]   Chen, B., Laverock, J., Piper, L. F. J., Preston, A. R. H., Cho, S. W., DeMasi, A., Smith, K. E., Scanlon, D. O., Watson, G. W., Egdell, R. G., Glans, P. –A., Guo, J. –H. *J. Phys.: Condens. Matter* 2013, **25**, 165501.

[188]   Kuzmin, A., Chaboy, J. *IUCrJ* 2014, **1**, 571.

[189]   Kuzmin, A., Anspoks, A., Kalinko, A., Timoshenko, J., Kalendarev, R. *Phys. Scr.* 2014, **89**, 044005.

[190]   Kuzmin, A., Anspoks, A., Kalinko, A., Timoshenko, J., Kalendarev, R. *Sol. Energ. Mater. Sol. Cell* 2015, **143**, 627.

[191]   Zhou, J., Wang, Y., Zhang, L., Li, X. *Chin. Chem. Lett.* 2017, *in press*.

[192]   Wieduwilt, T., Tuniz, A., Linzen, S., Goerke, S., Dellith, J., Hübner, U., Schmidt, M. A. *Sci. Rep.* 2015, **5**, 17060.

[193]   Guler, U., Boltasseva, A., Shalaev, V. M. *Science* 2014, **344**, 263.

[194]   Creighton, J. A., Eadont, D. G. *J. Chem. Soc. Faraday Trans.* 1991, **87**, 3881.

[195]   Cortie, M. B., McDonagh, A. M. *Chem. Rev.* 2011, **111**, 3713.

[196]   Zhang, X., Chen, Y. L., Liu, R. S., Tsai, D. P. *Rep. Prog. Phys.* 2013, **76**, 046401.

[197]   Harris, N., Arnold, M. D., Blaber, M. G., Ford, M. J. *J. Phys. Chem. C* 2009, **113**, 2784.

[198]   Atay, T., Song, J. H., Nurmikko, A. V. *Nano Lett.* 2004, **4**, 1627.

[199]   Zhong, Z., Pakovskyy, S., Bouvrette, P., Luong, J. H. T., Gedanken, A. *J. Phys. Chem. B* 2004, **108**, 446.

[200]   Thomas, K. G., Barazzouk, S., Ipe, B. I., Joseph, S. T. S., Kamat, P. V. *J. Phys. Chem. B* 2004, **108**, 13066.

[201]   Zhang, P., Wang, T., Gong, J. L. *Adv. Mater.* 2015, **27**, 5328.

[202]   Cushing, S. K., Li, J., Meng, F., Senty, T. R., Suri, S., Zhi, M., Li, M., Bristow, A. D., Wu, N. *J. Am. Chem. Soc.* 2012, **134**, 15033.

[203]   Li, J. T., Cushing, S. K., Zheng, P., Senty, T., Meng, F., Bristow, A. D., Manivannan, A., Wu, N. *J. Am. Chem. Soc.* 2014, **136**, 8438.

[204]   Fowler, R. H. *Phys. Rev.* 1931, **38**, 45.

[205]   Sze, S. M., Ng, K. K. *Physics of Semiconductor Devices*, 3rd Ed., John Wiley & Sons, Hoboken, NJ (2007).

[206]   Hölzl, J., Schulte, F. K., Wagner, H. *Solid Surface Physics*, 1st Ed., Springer, Berlin (1979).

[207]   Scaife, D. E. *Sol. Energy* 1980, **25**, 41.

[208]   Ingram, D. B., Linic, S. *J. Am. Chem. Soc.* 2011, **133**, 5202.

[209]   Kim, T. W., Choi, K. -S. *Science* 2014, **343**, 990.

[210]   Park, H. S., Lee, H. C., Leonard, K. C., Liu, G., Bard, A. J. *ChemPhysChem* 2013, **14**, 2277.

[211]   Ueno, K., Oshikiri, T., Misawa, H. *ChemPhysChem* 2016, **17**, 199.

[212]   Kim, H. -I., Monllor-Satoca, D., Kim, W., Choi, W. *Energy Environ. Sci.* 2015, **8**, 247.

[213]   Salvador, P. *J. Phys. Chem.* 1985, **89**, 3863.

[214]   Hisatomi, T., Kubota J., Domen, K. *Chem. Soc. Rev.* 2014, 43, 7520.

[215]   Jones, J. E., Hansen, L. D., Jones, S. E., Shelton, D. S., Thorne, J. M. *J. Phys. Chem.* 1995, **99**, 6973.

[216]   Bard, A. J., Faulkner, L. R. *Electrochemical Methods Fundamentals and Applications*, 2nd Ed., John Wiley & Sons, Hoboken, NJ (2001).

[217]   Hsu, Y. K., Chen, Y. C., Lin, Y. G., Chen, L. C., Chen, K. H. *J. Mater. Chem.* 2012, **22**, 2733.

[218]   Tachibana, Y., Vayssieres, Y., Durrant, J. R. *Nature Photon.* 2012, **6**, 511.

[219]   Hirakawa, T., Kamat, P. V. *J. Am. Chem. Soc.* 2005, **127**, 3928.

[220]   Jakob, M., Levanon, H., Kamat, P. V. *Nano Lett.* 2003, **3**, 353.

[221]   Thompson, T. L., Yates, J. T. Jr. *J. Phys. Chem. B* 2005, **109**, 18230.

[222]   Biagioni, P., Celebrano, M., Savoini, M., Grancini, G., Brida, D., Mátéfi-Tempfli, S., Mátéfi-Tempfli, M., Duò, L., Hecht, B., Cerullo, G., Finazzi1, M. *Phys. Rev. B* 2009, **80**, 045411.

[223]   Peter, L. M., Wijayantha, K. G. U., Tahir, A. A. *Faraday Discuss.* 2012, **155**, 309.

[224]   Kauranen, M., Zayats, A. V. *Nature Photon.* 2012, **6**, 737.

[225]   Khurgin, J. B. *Nature Nanotech.* 2015, **10**, 2.

[226]   Shalaev, V. M., Douketis, C., Stuckless, J. T., Moskovits, M. *Phys. Rev. B* 1996, 53, 11388.

[227]   Knight, M. W., Wang, Y., Urban, A. S., Sobhani, A., Zheng, B. Y., Nordlander, P., Halas, N. J. *Nano Lett.* 2013, **13**, 1687.

[228]   Clavero, C. *Nature Photon.* 2014, **8**, 95.

[229]   Hung, S. F., Xiao, F. X., Hsu, Y. Y., Suen, N. T., Yang, H. B., Chen, H. M., Liu, B. *Adv. Energy. Mater.* 2016, **6**, 1501339.

[230]   Furube, A., Du, L., Hara, K., Katoh, R., Tachiya, M. *J. Am. Chem. Soc.* 2012, **129**, 14852.

[231]   Bernardi, M., Mustafa, J., Neaton, J. B., Louie, S. G. *Nature Commun.* 2015, **6**, 7044.

[232]   Nishijima, Y., Ueno, K., Kotake, Y., Murakoshi, K., Inoue, H., Misawa, H. *J. Phys. Chem. Lett.* 2012, **3**, 1248.

[233]   Jiang, X. F., Pan, Y., Jiang, C., Zhao, T., Yuan, P., Venkatesan, T., Xu, Q. H. *J. Phys. Chem. Lett.* 2013, **4**, 1634.

[234]   Valenti, M., Venugopal, A., Tordera, D., Jonsson, M. P., Biskos, G., Schmidt-Ott, A., Smith, W. A., *ACS Photonics* 2017, **4**, 1146.

[235]   Chen, Z., Jaramillo, T. F., Deutsch, T. G., Kleiman-Shwarsctein, A., Forman, A. J.,
          Gaillard, N., Garland, R., Takanabe, K., Heske, C., Sunkara, M., McFarland, E. W.,
          Domen, K., Miller, E. L., Turner, J. A., Dinh, H. N. *J. Mater. Res.* 2010, **25**, 3.

[236]   American Society of Testing and Materials Standard: G173, 2003e1, in *Standard Tables
          for Reference Solar Spectral Irradiances: Direct Normal and Hemispherical on 37° Tilted
          Surface* (American Society of Testing and Materials International, West Coshohocken, PA,
          2003).

9. Appendix

9.1 List of Symbols and Abbreviations

A	Electron affinity
Ag/AgCl	Silver/silver chloride reference electrode
A_{ph}	Absorbance
ΔA_{ph}	Difference in abosrbance
$AuCl_4^-$	Gold tetrachloride anion
B	Full-width at half maximum of the diffraction peak
BD	Butanediol
BET	Brunauer–Emmett–Teller
BSE	Back-scattered electron
b	Path length
C	Coulomb
CB	Conduction band
CBED	Convergent-beam electron diffraction
C_{dye}	Concentration of organic dye
$C_{reac.}$	Concentration of the reactant
CRT	Cathode ray tube
C_s	Adsorption capacity
$C_{\beta\text{-}SnWO_4}$	Concentration of $\beta\text{-}SnWO_4$
$\nabla C_{\beta\text{-}SnWO_4}$	Concentration gradient of $\beta\text{-}SnWO_4$
c	Velocity of light
$c\text{-}SnWO_4$	$\beta\text{-}SnWO_4$ microcube
D	Diffusion coefficient
DEG	Diethylene glycol
DET	Direct electron transfer
DFT	Density functional theory
DLS	Dynamic light scattering
$D_o(E)$	Distribution of the energy states of electron acceptor O
DOS	Density of states
d	Diameter/hydrodynamic diameter
d_b	Base diameter
d_c	Cube size
d_{hkl}	Interplanar distance of the lattice plane (hkl)
d_{min}	Minimum distance
E	Energy

129

Appendix

ΔE	Energy change
ΔE_0	Correction of the energetic position of the X-ray absorption edge
E_{CB}	Conduction band edge of the semiconductor
EDXS	Energy dispersive X-ray spectroscopy
E_F	Fermi level
$E_{F,sc}$	Fermi level of the semiconductor
$E_{F,m}$	Fermi level of metallic counter electrode
$E_{F, redox}$	Fermi level of the redox couple
EG	Ethylene glycol
E_g	Bandgap of the semiconductor
ET	Everhart-Thornley detector
E_{VB}	Valence band edge of the semiconductor
$E_{V,SEI}$	Valence band edge at the semiconductor–electrolyte interface
EXAFS	Extended X-ray absorption fine structure
e	Base of natural logarithms
e^-	Electron
$e^-_{bulk,m}$	Electron in the bulk of the metal
$e^-_{bulk,sc}$	Electron in the bulk of the semiconductor
e_{CB}^-	Electron in the conduction band of the semiconductor
$[e_{CB}^-]$	Concentration of the electron in the conduction band of the semiconductor
$e^-_{depletion\ layer}$	Electron at the depletion layer of the semiconductor
$e^-E_{decomp.}$	Fermi energy of cathodic reduction of the semiconductor
e^-_{sc}	Electron of the semiconductor
e_{sp}^-	Electron in the sp-band of the metal
$[e_{sp}^-]$	Concentration of the electron in the sp-band of the metal
F	Faraday constant
fcc	Face-centered cubic
$F_{h\nu}$	Incident photon flux
$F_{\hbar\omega}$	Phonon flux
FMS	Full multiple scattering
$F(R_\infty)$	Kubelka-Munk function
FT	Fourier transforms
FWHM	Full-width at half maximum
G	Gibbs free energy
ΔG	Gibbs free energy change
GC	Gas chromatography
GLY	Glycerol
$\mathbf{g}$	Diffraction vectors

H	Height
HAADF-STEM	High-angle annular dark field-scanning transmission electron microscopy
HFM	Horizontal focusing mirror
$[H_2O]$	Concentration of water
HOMO	Highest occupied energy level/Highest occupied molecular orbital
HRTEM	High-resolution transmission electron microscope
h	Planck constant
$\hbar$	Reduced Planck constant
h^+	Hole
$h^+_{bulk,sc}$	Hole in the bulk of the semiconductor
h_d^+	Hole in the d-band of the metal
$[h_d^+]$	Concentration of the hole in the d-band of the metal
$h^+_{depletion\ layer}$	Hole at the depletion layer of semiconductor
h_{df}	Depth of filed
$h^+E_{decomp.}$	Fermi energy of anodic oxidation of the semiconductor
h_l	Arm length
h^+_{sc}	Hole of the semiconductor
$h^+_{semiconductor\text{-}liquid\ junction}$	Hole at the semiconductor-liquid junction
h_{sp}	Hole in the sp-band of the metal
$[h_{sp}^+]$	Concentration of the hole in the sp-band of the metal
h_{VB}^+	Hole in the valence band of the semiconductor
$[h_{VB}^+]$	Concentration of the hole in the valence band of the semiconductor
$h\nu$	Photon energy
$\hbar\omega$	Phonon energy
I	Ionization potential
I_0	Radiant intensity of incident light/X-ray
ICDD	International center for diffraction data
I_{in}	Initial photocurrent shoot at the light-on instant
IPCE	Incident photon to charge carrier efficiency
I_{ph}	Photocurrent density
I_s	Radiant intensity of scattered light
I_t	Radiant intensity of transmitted light/X-ray
ITO	Tin-doped indium oxide
J	Diffusion flux
$K/\alpha/\varepsilon_{molar}$	Absorption coefficient
K_{ad}	Adsorption equilibrium constant
k	Wave number
Δk	Momentum change

Appendix

k'	Photodegradation rate constant
k_B	Boltzmann constant
kg	Kilogram
k_i	Rate constant of ith reaction
k_r	Charge recombination rate constant
k_{red}	Interfacial reaction rate constant
k-vector	Wavevector of the photon
L	Distance from the specimen to the photographic negative in the TEM
L_e	Mean free diffusion length of the electron
L_h	Mean free diffusion length of the hole
LSP	Localized surface plasmon
LSPR	Localized surface plasmon resonance
LUMO	Lowest unoccupied energy level/(Lowest unoccupied molecular orbital
l	Orbital angular momentum quantum number
Δl	Change of orbital angular momentum quantum number
MB	Methylene blue
MFP	Mean free path
m	Mass
m_e	Rest mass of the electron
N	Coordination number
N_D	Charge carrier density of the n-type semiconductor
NHE	Normal Hydrogen Electrode
NIR	Near-infrared
NSRRC	National synchrotron radiation research center
n	Relative refractive index
n_b	Bulk density of electrons
n_d	Order of diffraction
n_{molar}	Mole
n_s	Surface density of electrons
O/O'	Electron acceptors in the electrolyte
$[O]$	Concentration of electron acceptor O in the electrolyte
OAc	Oleic acid
OAm	Oleylamine
PDO	Propanediol
PE	Pentaerythritol
PEG	Polyethylene glycol
P_{in}	Power density of incident light
p	Momentum

pDOS	Projected density of states
QD	Quantum dot
Q_{ph}	Photogenerated charge
q	Electronic charge
$q\Phi_b$	Schottky barrier height
R/R'	Electron donors in the electrolyte
$[R]$	Concentration of electron donor R in the electrolyte
R_{bond}	Bond length
RET	Resonant energy transfer
$Re\chi(R)$	Real part of normalized oscillatory part of the X-ray absorption coefficient with cubed weighting of the wave number of X-ray
R_f	Goodness-of-fits
RhB	Rhodamine B
R_p	Distance of incident light to the particle
$R_{reference}$	Reflectance of the reference standard
R_{sample}	Reflectance of the sample
R_∞	Relative reflectance
Rxn	Reaction
r	Radius
r_{ED}	Distance from the diffracted to the transmitted spots on the electron diffraction pattern
r_{ph}	Photodegradation rate of organic dye
S	Scattering coefficient
S_0^2	Amplitude reduction factor of oscillatory part of the X-ray absorption coefficient
SAED	Selected area electron diffraction
SE	Secondary electron
SEM	Scanning electron microscopy/microscope
SOJT	Second-order Jahn-Teller
SPP	Surface plasmon polariton
SS	Mid-gap surface state
$[SS]$	Density of the mid-gap surface state
SS^+	Hole at the mid-gap surface state
s-SnWO$_4$	β-SnWO$_4$ spikecube
STEM	Scanning transmission electron microscopy
T	Absolute temperature
TBAOH	Tetrabutylammonium hydroxide
TBAOH·30H$_2$O	Tetrabutylammonium hydroxide 30-hydrate

TCD	Thermal conductivity detector
TEM	Transmission electron microscopy/microscope
TEY	Total electron yield
TFY	Total fluorescence yield
TOF	Turnover frequency
TON	Turnover number
T_{ph}	Transmittance
TrEG	Triethylene glycol
t	Time
t_c	Diameter of the crystallite
t_s	Specimen thickness
UHV	Ultrahigh vacuum
ΔU_p	Underpotential shift
UPD	Underpotential deposition
UV	Ultraviolet
V	Accelerating voltage
VB	Valence band
V_B	Extent of band bending
V_{fb}	Flatband potential
VFM	Vertical focusing mirror
V°_{redox}	Standard redox potential of the redox couple
VIS	Visible
v	Velocity
w	Width of the depletion layer
XAFS	X-ray absorption fine structure
XANES	X-ray absorption near edge structure
XAS	X-ray absorption spectroscopy
XRD	X-ray powder diffraction
$\overline{(\Delta x)}^2$	Mean squared displacement
Z	Atomic number
z	Number of transferred charge
ZB	Zero-order beam
(hkl)	Miller indices of the lattice plane
$\{hkl\}$	Miller indices of a family of the lattice planes
Γ	Core-hole lifetime broadening
$\overline{\mu_e}$	Chemical potential of the electron
$\mu(E)$	X-ray absorption coefficient
λ	Wavelength

λL	Camera constant
λ_s	Solvent reorganization energy
ν	Frequency of incident photon
ω	Frequency of the harmonic oscillator lattice
ε	Dielectric constant of the semiconductor
ε_0	Permittivity of free space
α^{-1}	Light penetration depth
α_{con}	Semi-angle of radiation cone/convergence angle
χ	Electronegativity
$\chi(k)$	Normalized oscillatory part of the X-ray absorption coefficient
θ	Scattering angle
2θ	Diffraction angle
θ_B	Bragg angle
θ_s	Surface coverage
η	Viscosity
η_F	Faradaic efficiency
η_i	Internal quantum transmission probability
δ	Partial electron transfer
Φ	Work function
$\Delta\Phi$	Work function difference
Φ_{ph}	Photoabsorption efficiency
σ^2	Debye-Waller factor
σ	Distribution width of the bond length
τ_{sp}	Relaxation time of the hole in the sp-band of the metal
τ_d	Relaxation time of the hole in the d-band of the metal

9.2 List of Figures

9.3 List of Tables

10. Curriculum Vitae

Personal

Name: Ying-Chu Chen
Date of birth: 28.03.1987
Place of birth: Taipei, Taiwan
Citizenship: Taiwanese

Academic Qualifications

10/2013-02/2018	Ph. D. in Inorganic Chemistry
	Karlsruhe Institute of Technology, Germany
Thesis title:	*Advanced Nanoarchitectures for Photocatalytic Functionality*
09/2011-01/2013	M. Sc. in Chemical Engineering
	National Taiwan University, Taiwan
Thesis title:	*Nanostructured Iron Oxyhydroxide (FeOOH) as a Novel Anode Material for Asymmetric Electrochemical Capacitor*
09/2005-06/2009	B. Sc. in Chemical Engineering
	National Cheng-Kung University, Taiwan
09/2002-06/2005	High school education
	Taipei Municipal Datong High School, Taiwan

Research Experiences

02/2013-05/2013	Research Assistant
	Center for Condensed Matter Sciences
	National Taiwan University, Taiwan
06/2009-09/2011	Research Assistant
	Center for Condensed Matter Sciences
	National Taiwan University, Taiwan

11. List of Publications

Academic papers

1. Chen, Y. C., Hsu, Y. K., Popescu, R., Gerthsen, D., Lin, Y. G., Feldmann, C. *Nature Commun.* 2018, **9**, 232.
2. Chen, Y. C., Hsu, L. C., Lin, Y. G., Tarasov, A., Chen, P. T., Hayashi, M., Ungelenk, J., Hsu, Y. K., Feldmann, C. *ACS Catal.* 2016, **6**, 2357.
3. Dong, H., Chen, Y. C., Feldmann, C. *Green Chem.* 2015, **17**, 4107.

International Conference

1. Chen, Y. C., Popescu, R., Gerthsen, D., Hsu, Y. K., Lin, Y. G., Feldmann, C. Poster P22, **13th International Conference on Materials Chemistry**, 10.07.2017-13.07.2017, Liverpool, United Kingdom.
2. Chen, Y. C., Feldmann, C. Poster P069 **18. Vortragstagung Fachgruppe Festkorperchemie und Materialforschung**, 19.09.2016-21.09.2016, Innsbruck, Austria.
3. Chen, Y. C., Ungelenk, J., Hsu, Y. K., Lin, Y. G., Feldmann, C. Poster P136, **12th International Conference on Materials Chemistry**, 20.07.2015-23.07.2015, York, United Kingdom.
4. Chen, Y. C., Feldmann, C. Poster P136, **17. Vortragstagung Fachgruppe Festkorperchemie und Materialforschung**, 15.09.2014-17.09.2014, Dresden, Germany.

12. Acknowledgement

First and foremost I would like to express my sincere gratitude to my advisor, Prof. Dr. Claus Feldmann, for giving me the opportunity to study nanomaterials chemistry and for support throughout this work as he provided many insightful ideas and thoughts during my graduate career. His academic guidance, constant encouragement, patience, time, and enthusiasm were highly appreciated. In addition, I am grateful to him for the tremendous effort that went into each and every draft of my dissertation.

Moreover, I gratefully acknowledge Prof. Yu-Kuei Hsu for his support, fruitful cooperation and helpful discussions throughout my PhD. He constantly pushed the quality of my research towards excellence. He was a great mentor and will continue to be a friend for many years.

I would like to extend my appreciation to Prof. Dr. Dagmar Gerthsen, Dr. Radian Popescu and Dr. Yan-Gu Lin for their great efforts in performing excellent TEM and XAS measurements and carefully evaluating associated materials, respectively.

I would also like to thank the entire Feldmann research group for their help and kindness throughout these years. Financial support from the scholarship (No. A/13/92805) of Deutscher Akademischer Austauschdienst (DAAD) is likewise highly acknowledged.

Last but not least, I would like to thank my families for their endless love and encouragement that are absolutely essential.

www.ingramcontent.com/pod-product-compliance
Lightning Source LLC
Chambersburg PA
CBHW070728220326
41598CB00024BA/3350